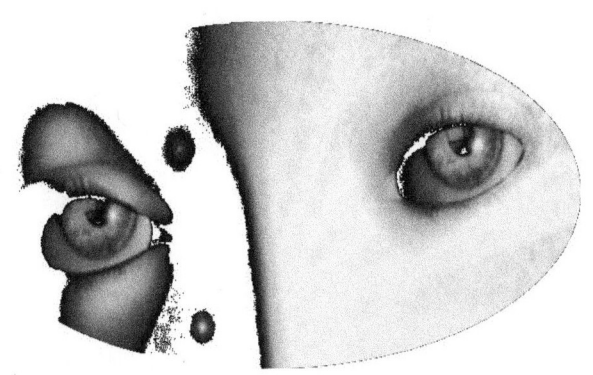

Nightfall's Day

by Lyn Murray

Golden Panda Publishing
USA

Copyright

Introduction

. . . What you'll find in this anthology . . .

Some of what is enclosed are "actual files" (collected and compiled just for you and left as original as possible) that translated beautifully into print, so you might want to pick up a hardcopy of this book to keep (while sharing the eBook version) if you are a lover and collector of weird historical facts that have a ***tendency*** to disappear over time.

Why the ***tendency*** prevails is a short stretch of the imagination for those of us who refuse to remain Sheeple (i.e. doing what we're told when we're told to do it – abandoning freedom of thought and action) which is exactly why *YOU* have chosen to pick up this book!

Welcome – ***Free Thinkers*** – Everywhere!

!! Come On In !!

Nightfall's Day

What if Time Isn't Time?

by Lyn Murray

Dedication

To my family for putting up with me.
To my fans for being faithful!
To those who experience strange occurrences
– where would I be *without* you.

Foreword

Not all roads should be traveled.

But, what if the road you take is not up to you; and you're the only one who knows you've just driven down a nightmare?

What could be more innocent than a harmless day-trip in the country? Practically nothing; unless you venture down the path less traveled – and end up driving headlong into a nightmare that alters reality, plunging you into Nightfall's Day, a day without end, a day without a night; and a nightmare from which you will never awaken!

Watch your speed!

Preface

Nightfall's Day is a study in (and of) TIME, combining fact filled discussions and fascinating stories that are based upon truths you may or may not be aware of.

We all assume that time is the same everywhere – when that couldn't be further from the truth! Time is not a constant, and changes when speed/acceleration is part of the equation. For instance – if you live in a two story house, time is *marginally (but measurably) slower* upstairs than it is downstairs.

What if scientists have discovered ways to create time anomalies within Earth's gravitational field that can alter time, change the fabric of time, and maneuver time to facilitate some hidden agenda? *What if –* we're caught in the middle of this technology, and only realize it when we stumble into and are trapped in a time anomaly. Could this be one explanation for *DejaVu?* What are you going to do? *Die?!* Or – *perhaps* find yourself lost between dimensions – *forever.*

Fantasy? *Maybe not –* as you'll discover when reading about the electromagnetic warfare involving *HAARP – that's going on now, behind our backs, right before our sheeple eyes!*

Table of Contents

HAARP

High Frequency Active Auroral Research Program

**"What you choose to believe –
is yours to decide!"**

The Fourth Kind

What is HAARP?

High Frequency Active Auroral Research Program

This is an ionospheric research program jointly funded by the U.S. Air Force, the U.S. Navy, the University of Alaska, and the Defense Advanced Research Projects Agency (DARPA). Designed and built by BAE Advanced Technologies (BAEAT), its purpose is to analyze the ionosphere and investigate the potential for developing ionosphere enhancement technology for radio communications and surveillance. The HAARP program operates a major sub-arctic facility, named the HAARP Research Station, on an Air Force–owned site near Gakona, Alaska.

The most prominent instrument at the HAARP Station is the Ionospheric Research Instrument (IRI), a high-power radio frequency transmitter facility operating in the high frequency (HF) band. The IRI is used to temporarily excite a limited area of the Ionosphere. Other instruments, such as a VHF and a UHF radar, a fluxgate magnetometer, a DiGisonde (an ionospheric sounding device), and an induction magnetometer, are used to study the physical processes that occur in the excited region.

Work on the HAARP Station began in 1993. The current working IRI was completed in 2007, and its prime contractor was BAE Systems Advanced Technologies. As of 2008, HAARP had incurred around $250 million in tax-funded construction and operating costs. It was reported to be temporarily shut down in May 2013, awaiting a change of contractors. As of May 2014, it has been announced that the HAARP program will be shut down later in the year – (*but they've said this before*).

HAARP is a target of conspiracy theorists, who claim that it is capable of modifying weather, disabling satellites and exerting mind control over people and that it is being used as a weapon against terrorists. Such theorists have blamed the program for causing earthquakes, droughts, storms and floods, diseases such as Gulf War Syndrome and Chronic Fatigue Syndrome, the 1996 crash of TWA Flight 800, and the 2003 destruction of the space shuttle *Columbia*. Commentators and scientists say that proponents of these theories are "uninformed", as most theories put forward fall well outside the abilities of the facility and often outside the scope of natural science.

(*No one ever said HAARP was involved in Natural Science!*)

HAARP

Fact Sheet

HAARP (**H**igh Frequency **A**ctive **A**uroral **R**esearch **P**rogram) is to be a major Arctic facility for upper atmospheric and solar-terrestrial research. HAARP is built on a DoD-owned site near Gakona, Alaska. Principal instruments include a high power, high-frequency (HF) phased array radio transmitter (known as the Ionospheric Research Instrument, or IRI), used to stimulate small, well-defined volumes of ionosphere, and an ultra-high frequency (UHF) incoherent scatter radar (ISR), used to measure electron densities, electron and ion temperatures, and Doppler velocities in the stimulated region and in the natural ionosphere. To further the scientific capabilities and usefulness of the IRI and ISR, HAARP is supporting the design and installation of the latest in modern geophysical research instruments, including an HF ionosonde, ELF and VLF receivers, magnetometers, riometers, a LIDAR (Light Detection And Ranging) and optical and infrared spectrometers and cameras which will be used to observe the complex natural variations of Alaska's ionosphere as well as to detect artificial effects produced by the IRI.

HAARP

BOILS THE UPPER ATMOSPHERE –

HAARP will zap the upper atmosphere with a focused and steerable electromagnetic beam. It is an advanced model of an 'ionospheric heater'. (The ionosphere is the electrically-charged sphere surrounding Earth's upper atmosphere. It ranges between about 40 to 600 miles above Earth's surface.) Put simply, the apparatus for HAARP is a reversal of a radio telescope: antennas send out signals instead of receiving. HAARP is the test run for a super-powerful radio wave beaming technology that lifts areas of the ionosphere by focusing a beam and heating those areas. Electromagnetic waves then bounce back onto Earth and penetrate everything-living and dead. HAARP publicity gives the impression that the High-frequency Active Auroral Research Program is mainly an academic project with the goal of changing the ionosphere to improve communications for our own good. However, other US military documents put it more clearly: HAARP aims to learn how to "exploit the ionosphere for Department of Defense purposes".

HAARP: Vandalism in the Sky?

We are assured . . . *by everyone from Ph.D.s to poets* that HAARP is:

1. An ionospheric heater

2. A research tool

3. A military test bed

4. A tax-subsidized boondoggle

5. A directed-energy weapon

6. A communication system for submarines

7. A source of field-aligned ionospheric VHF reflectors

8. A way to improve satellite links

9. A planetary x-ray machine

10. A plot to depopulate the Third World

11. A means of creating power blackouts at will

12. Electronic warfare

13. Tesla's wireless power transmission

14. Tesla's secret death ray

15. Searching for space aliens

16. Killing space aliens

17. Killing off the militias

18. Keeping them awake at night (through RF head rectification)

19. Enforcing the New World Order

20. Creating nuclear-scale explosions

21. Weather modification

22. CIA mind control

23. Brainwave modification

24. The end of HF radio

25. The end of wildlife in Alaska

26. The end of atmospheric ozone

27. The end of the human race

28. The end of Earth itself.

You must admit, that's pretty good for one transmitter. Not since the one built by

Marconi – has an experimental radio drawn such attention. *However, where there is smoke – there is usually fire!*

So – What's really up with HAARP?

◇◇◇

HAARP
Conspiracy

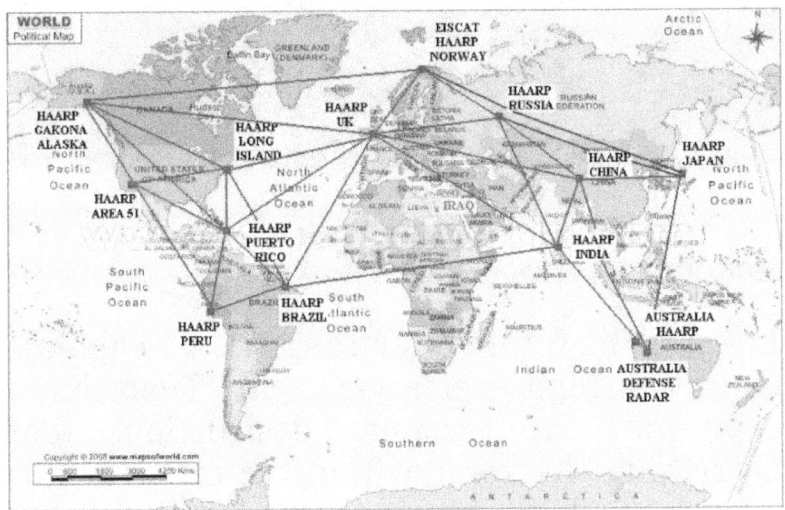

◇◇◇

New World Order
Mind Control
&
Weather Warfare

◇◇◇

Electromagnetic Warfare

The world of electromagnetic warfare and weapons of mass destruction

Imagine the future – a strange new weapon is detonated high over a large city. There is no explosion, no visible destruction, but everything electronic within the range of this weapon will go out – permanently. Every electronic gadget in every home and office – disabled. No computers, no T.V., no life support systems in hospitals, no water supply, no heat, no lights – truly, a return to the dark ages. Imagine a full range of new weapons; one can take out the electricity in your city, another can destroy you. If you haven't heard about these weapons, it's no surprise. Their development has been secretive, and they sound more like science fiction than reality. *When did this reality really begin and how far advanced is it now?*

On a calm Sunday morning in 1978, residents of Bell Island, Newfoundland hear an odd, high-pitched hum, immediately followed by a sudden and terrifying blast resounding for hundreds of miles. Witnesses observe a "straight shaft of light" descending from the sky and a barrage of violent electrical phenomenon. Outbuildings are destroyed, livestock electrocuted, televisions explode and power lines vaporize. Observed from space, the light emissions from the "boom" are more powerful than those of the Hiroshima blast.

The documentary: The Invisible Machine unravels the mystery of the Bell Island "boom" and in doing so takes a chilling look at the U.S. military's experimentation with electromagnetic pulse (EMP) weapons, "e-bombs", directed energy weapons that can destroy electrical and communications systems but leave people seemingly unharmed. Reporters interviewed eye-witnesses, an EMP weapons designer, scientists, journalists, policymakers and activists who warn that these weapons are real and may have been used by the US military during its "shock and awe" campaign against Iraq in 2003.

While the United States continues to develop and experiment with these weapons of mass destruction, the questions remain: were the bizarre events on Bell Island in 1978 related to early testing gone wrong and were directed EMP weapons at Baghdad's electricity generation plant the most recent experiment? If so, what are the consequences of unleashing this powerful force? *The Invisible Machine* pulls back the veils of secrecy to find the answers.

The Tunguska Explosion

On June 30, 1908, a comet fragment exploded in Earth's atmosphere in Russian Siberia. A great blue-white fireball, brighter than the Sun streaked through the sky, exploded six to eight kilometers in the atmosphere with a blinding flash and intense pulse of heat; with electromagnetic pulse like anomalies were reported. The magnetic storm began a few minutes after the explosion. A compass was useless in Irkutsk, 1,000 km away. The explosion was heard 1,000 km away with trees were flattened 30 km from a central point in the Stony Tunguska River Valley. Sides of trees were burned 60 km away. The blast destroyed over 600 square kilometers of forest as the pillar of smoke and dust rose over the area. No crater was found. Scientists have observed the destruction and concluded the explosion had the force of a 30-megaton hydrogen bomb.

1 mile = 1.609344 Kilometer

WEATHER WARFARE
A Closer Look at HAARP's Dark Side

Here we'll take a basic look at the technology involved in controlling the weather, and try to KISS it (keep it simple) so that you can understand a most complex scientific subject.

The New World Order is coming!

Are you ready?

Once you understand what this New World Order really is, and how it is being gradually implemented, you will be able to see it progressing in your daily news!!

Learn how to protect yourself, your loved ones *and the Earth herself.*

Now for insights so startling you will never look at the news the same way again:

The idea behind Weather Control is simple, when you think about it simply. When you see and experience a strong thunderstorm, with a lot of lightening and thunder, what fact about this storm strikes you the most? Are you not impressed by the powerful display of energy that you witness?

Energy is the primary ingredient behind nature's storms. Therefore, you must believe that (just perhaps) if energy is the most dominant outward factor in all kinds of storms, then *energy might be the key factor in creating such storms in the first place.*

ENERGY, ENERGY, ENERGY, ENERGY, and More ENERGY!

So, you ask, how much energy is required to create, and then direct, storms? The answer to that question depends upon many factors, but let me tell you how much capability has been built into the newly created power transmission station in remote Alaska. These power transmission towers are not your typical towers, as they are designed to generate power in such a way that it is beamed up into the ionosphere in tremendous quantities.

"The $30 million [Pentagon] project, euphemistically named HAARP (High-Frequency Active Auroral Research Program), is made to beam more than 1.7 gigawatts (billion watts) of radiated power into the ionosphere – the electrically charged layer above Earth's atmosphere. Put simply – the apparatus is a reversal of a radio telescope – just transmitting instead of receiving. It will *'boil the upper atmosphere'. After [heating] and disturbing the ionosphere, the radiations will bounce back onto the earth in the form of long waves which penetrate our bodies, the ground, and the oceans."*

"Angels Don't Play This HAARP!"

By

Dr. Nick Begich and Jeane Manning

Dr. Begich explains this concept. "This invention provides the ability to put unprecedented amounts of power in the Earth's atmosphere at strategic locations and to maintain the power injection level, particularly if ***random pulsing*** is employed, in a manner far more precise and better controlled than heretofore accomplished by the prior art."

The goal is to learn how to manipulate the ionosphere on a grander scale than the Soviet Union could do with its similar facilities. **_HAARP would be the largest ionospheric heater in the world,_** located in a latitude most conducive to putting Eastlund's invention into practice." Furthermore, from this northern latitude, the energy could be aimed into the ionosphere so that it would bounce back down to the earth so it would come down wherever the scientists wanted it to come down. The secret was to learn how and where to aim it to hit the earth where they wanted it to hit, creating the type of disaster or weather they desired.

In a nutshell, this is the nucleus of the expertise just recently acquired to control the weather. By pouring measured energy that has been focused into certain parts of the ionosphere, scientists can create all kinds of storms, like **_hurricanes, thunderstorms, floods, tornadoes, and drought_**. In NEWS 1198 "U.N. Treaty Proves Weather Control Is Real", we report news articles that Malaysia actually contracted with a Russian Weather Modification company to create a hurricane that would be directed close enough to clear the smoke and smog from Malaysia's cities without actually coming on to land to create devastation. This Russian company delivered, and Malaysia had clear skies.

Our information also tells us that, not only can hurricanes be created, they can be dismantled should scientists so desire. And – they certainly can be driven on the ocean much like we drive our cars on roadways. Therefore, one has to ask why American scientists have allowed unprecedented hurricanes, like Andrew, ever to come on shore. Why are American scientists allowing extensive damage and lives lost to recent unprecedented storms since they have the capability to keep these storms away from us?"

Doesn't our own American Government have our best interests at heart?

Keep that thought in mind as we examine still more aspects of this HAARP technology that is pouring such enormous quantities of energy into our upper atmosphere. Researches quickly found that this technology could be used in **ways other than just to control the weather**. They discovered they had stumbled upon a weapon which could be used most effectively, to **destroy, destroy, and destroy** some more, with the vast majority of the peoples of the world completely unaware of what was happening to them. *After all, most people today still believe that the control of the basic weather of this planet is out of the control of mankind.*

Christians believe only God can control the weather, and they take great comfort in this belief. After all, we know that Earth is a small planet hurtling through an empty, cold, and inhospitable space at over 60,000 miles per hour. Most of us instinctively know that we can only trust an Omnipotent and All-Wise God to control the basic operating systems of planet earth. If we even thought that man, with all his inherent wickedness, could seize control of the basic operating systems of earth, and the entire earth could be ruined by man's deliberate wickedness or by his lack of understanding of the power he has now placed in his hand, we would see panic on a scale unimagined before! *Do not misunderstand: Earth could be destroyed, made unfit for human habitation, by mistakes accidentally created in the lab (it's been done before: Example Aids . . . and others!) That's why we need to adopt the "if it ain't broke – don't fix it policy" – and leave mother nature alone!*

New Age scientists of "Angels Don't Play This HAARP" are worried about just this type of accidental damage to earth. They are worried that the following damage might be done to this earth, some of which might render this planet uninhabitable. If you are a Christian that literally believes the Scriptures, you know that God will not allow this planet to become uninhabitable, as He will not allow man to completely destroy what God has created and declared good.

However, we also know that the Book of Revelation foretells unprecedented destruction that God either directs or allows, that "destroys the sinners out of the land" [Isaiah 13:9-13]. God judges wicked man at the same time as He is cleansing the earth to prepare for the arrival of Jesus Christ to begin his Millennial reign.

We are very interested in the fact that many of the "lying signs and wonders" of the Antichrist can be staged by the technology of HAARP. Example: ***Project Blue Beam.***

What is the Blue Beam Project?

It involves two things. A technologically simulated "second coming" and the reemergence of new "Montauk" type projects that have the ability to take up a whole bunch of people as in a "rapture" type of situation and whisk the whole bunch into never-never land.

Ironically, portions of the holographic projections have the potential for changing the planet into oneness with God. Unfortunately, this operates on the premise that Man shall somehow become God in human form and control other Men and dictate all actions and thoughts.

The calculated resistance to the new religion, the New World Order and the new "Messiah" will entail human loss on a massive scale in the ensuing *"holy wars"*. The "BLUE BEAM PROJECT" will pretend to be the universal fulfillment of the prophecies of old; as major an event as that which took place 2000 years ago. In principle, it will make use of the sky as a holographic projection screen for space-based laser-generating satellites (star wars). These projectors will project simultaneous images to the four corners of the planet, in every language by region.

Furthermore – many of the physical acts of judgment contained in the Book of Revelation, and parallel passages in the Major and Minor

Prophets, can be created by this HAARP technology.

Given the desperately wicked heart of man, perhaps the most damage God could do would be to allow man to get unprecedented ability and power in his grasp, and then move His Holy Spirit out of the way to allow man to treat other men as his wicked heart would direct!

- We find it very interesting that God's wording in the Book of Revelation seems to back up this interpretation that God may allow man to possess technological power that he could use against other men, according to the wickedness of his heart. Look at the wording of some of these passages. The word, "given", in each of these Scriptures means "to give", to "grant", to "receive". In other words, these Scriptures are saying that God is going to allow the men of Antichrist to receive abilities that they never possessed before. These abilities will then be used to bring God's Judgments upon this wicked earth and its rebellious people.
- **Revelation 6:4-5** – This passage foretells the next horse that rides out of Heaven after Antichrist rides,

symbolizing his appearance on the earth. "And there went out another horse that was red: and **power** was given to him that sat thereon to take peace from the earth, and that they should kill one another: and there was given unto him a *great* sword." In this instance, mankind will receive the ability to wage unprecedented warfare upon the inhabitants of the earth.

- **Revelation 6:8** – This passage foretells that man will be given unprecedented ability to kill other men with unprecedented means. "And I looked and beheld a pale horse: and his name that sat on him was Death, and Hell followed with him. And power was given unto them over the fourth part of the earth, to *kill with sword*, and with *hunger*, and with *death*, and with the *beasts of the earth*." We shall demonstrate that HAARP technology can wield this kind of weapon upon the peoples of the earth.

- **Revelation 13:5** – Antichrist is given unprecedented ability by God to reign with a degree of wickedness and audacity never seen in history. "And there was given unto him a mouth speaking great things and blasphemies, and power was given unto him to continue forty and two months." While HAARP cannot give the Antichrist such a great mouth, it can give him the ability to project his speeches all over the world, where people can see his image in the sky above them, speaking to them in their own language!

- **Revelation 13:7** – Antichrist is given the ability to slaughter the saints of God. "And it was given unto him to make war with the saints and to overcome them: and power was given him over all kindreds, and tongues, and nations." I find it highly instructive that the same unprecedented ability which God is going to allow Antichrist to rule mankind and slaughter the saints of God is the same type of unprecedented technological ability God is going to give man in these other passages.

- **Revelation 16:8-9** – HAARP can create this scourge of heat from the Sun! "And the fourth angel poured out his vial upon the sun, and power was given unto him to scorch men with fire. And men were scorched with great heat, and blasphemed the name of God, which hath power over these plagues." I find it highly instructive that, while the last part of this passage says God retains power over these plagues, the first part of the passage says that God is going to "give" unto wicked men unprecedented powers by which other men are going to be judged! In other words, while God is allowing Satan to reveal unprecedented ability to human scientists to inflict death and suffering on wicked man, God retains ultimate control. *Once created – a situation MUST play itself out!*

Now let us examine some of the concerns of these New Age scientists of the damage HAARP can inflict upon the earth. HAARP can:

- Create Earthquakes on demand wherever and whenever you wish on the earth. [Using HAARP to create earthquakes.] However, our military discovered, by accident, how to create earthquakes using slightly different technology. "The planetary scale engineers tried to replace a ten by forty kilometer section of the ionosphere with a 'telecommunications shield' of 350,000 copper needles tossed into orbit. When the military sent up a band of tiny copper wires into the ionosphere to orbit the planet so as to 'reflect radio waves and make reception clearer', we had the 8.5 Alaska earthquake and Chile lost a good deal of its coast. That band of copper wires interfered with the planetary magnetic field."

- Manipulate Global Weather Systems, changing weather patterns, rainfall, and drought.
- Earth going wild in its rotation, possibly spinning out of control.
- Redirecting Jet Streams.
- Redirecting flow of Electro Jet Stream.
- Magnifying and Focusing sunlight, called 'Sky busting'. This process could burn holes in the protective ozone layers, thus allowing intense sunlight to pour through, burning mankind severely. Look at Revelation 16:8, quoted above.
- Mind Control. Using extremely low-frequency bombardment at just the same frequency as the human brain operates, you can change a person's thoughts or emotions. God help us all when and if wicked men achieves finesse with this weapon!

HAARP can create nuclear-sized explosions without radiation!. This process is protected by patent 4,873,928. This might be the "great" sword being wielded by the forces of Antichrist in Revelation 6:4. With this weapon, you could attack targets with nuclear-sized explosions without having to deliver nuclear warheads on missiles, or on aircraft, or any carrier!! This development could render all military calculations about how to defend against an enemy's attack completely useless.

Whoa – mental snapshot! Now I understand why the Russian and American negotiators were suddenly able to reach agreement on the destruction of so many warheads and missiles. Each side even allowed inspection teams to oversee the destruction of actual warheads and missiles. *An illusionary tactic – but the world did feel much safer to the grunts of society.*

However, *warheads carried on missiles are obsolete!!* If both sides can create nuclear-sized explosions without radiation by HARRP-type radio transmitting towers, you would not need missiles to carry warheads. How can you defend against a nuclear-sized attack that is being initiated by silent, invisible ELF radio waves? How do you defend against a nuclear-sized attack when it is being initiated by ELF radio waves that are being generated on enemy territory, fly up to the ionosphere above his territory, and then bounce back down to your territory to create the explosion!

Clearly . . . the world has entered an entirely new world of warfare that the *grunts* of society never saw coming that I believe is being controlled by **The Dark Side**: the antichrist and his minion – for the sole purpose of controlling the earth and its occupants; and like heads of state around the world – scientists skip along happily, doing their bidding . . . sending the world into chaos!

- Furthermore, this subject brings us to the next capability of HAARP and like radio wave technology. These ELF radio transmitters can throw up an *impenetrable shield* against incoming missiles, aircraft, or anything trying to get through our airspace! Then, why are we attempting to perfect an anti-missile missile that can actually hit another missile in flight? Why is the American military moving rapidly in conjunction with the Israeli military to bring this more effective anti-missile missile to Israel so as to better protect against Syrian and Egyptian missiles? They probably don't want very many people knowing that we possess this type of effective missile shield over any battlefield, either strategic or tactical. When the Arabs engage the Israelis in this next war, they just might be shocked at the type of weapons that are annihilating them! The existence of this type of military capability is also the reason we are allowing the Arab governments to buy American weaponry to fight Israel in this next war. The Arabs don't understand that the Israeli military (and we) have taken a quantum leap forward in destructive, killing, and annihilating capability!

In fact, has the infantryman finally been made obsolete in warfare? *Obsolete yes – but their innocent blood feeds the flames of lust and greed!*

The greatest concern of New Age scientist Nick Begich, in his book, **"Angels Don't Play This HAARP"**, is that scientists and the military are so very arrogant in their ignorant, reckless use of focused energy into the ionosphere. Military scientists talk about "kicking this thing in high gear to see what would happen"! This attitude is complete arrogance. What if they set off an unintended reaction in the atmosphere that cannot be controlled or stopped? Indeed, I think some of the judgments foretold in the Book of Revelation might come from this type of major miscalculation by scientists.

The Creator is not the author of confusion, war, pestilence, and is not responsible for our ability and more than willing annihilation of each other and planet Earth; but the very fact that we have WMD (Weapons of Mass Destruction) them and use them tells me that we are near the end of our age . Are you ready to meet the One True God? Are your loved ones? Have you adequately protected your family?

Listen to me now: it is this author's belief that mankind's DNA was manipulated by an alien race: the Anunnaki (*whatever they're calling themselves these days*) – who cannot create the life force (only the Ultimate Creator can create life from nothing and give it substance). However, these beings, aliens, fallen angels, *alien scientists*, Satan – took what life they found here, rearranged that life's DNA – *and we were born!*

I also believe that there are *MANY* voices in the Bible (all religious text from around the world) and that our worship of the One True God has been corrupted by the aliens, demons, fallen angels, *alien scientists*, Satan, and they have deceived us and tricked us into thinking they are THE GOD – when they are nothing more than scientist creating life in a petri dish! *They gave us form – not life!*

Their corruption of that form is why The Living God (the entity we know as Jesus Christ) came here over 2,000 years ago – to tell us we were worshiping the wrong god, and to show us the correct path to follow.

It absolutely is!

If you've ever wondered why the Bible (*all religious text*) seems contradictory – now you know, and the reason is that there are more voices in there than that of our Creator.

Let's make this simple: it is my belief that wherever you see divisiveness, manipulation, exploitation (in all of it's many forms), jealousy, greed, death and destruction, war, sacrificial death (*yes, you heard right!*) – then you are NOT dealing with The One True God of Creation and himself in the person of his own Son, Jesus Christ.

You're NOT! You are dealing with a subordinate with a God Complex! You are!

This is WHY the creator came in the person of his own Son – Jesus – to set the record straight, and get us back on track and following him!

Still having trouble believing the Alien (Fallen Angel Demons) genetic engineering of the human race?

You're not alone!

Why?

Because people have trouble separating Genetic Creation from Life Force Creation . . . but there is a difference. *A Big One!*

Only the CREATOR of ALL THINGS can create the LIFE-FORCE that gives the genetic creation life.

Believe it!

Now that we're past that hurdle . . . consider this . . . :

   If WE (the human race) – were doing it right . . . why did we need a Savior?

The one and only reason He came – was to tell us that we were doing it wrong! And when he came – everything changed.

Everything!

Perfect LOVE!!

Please try to understand this, because the evil fallen angels are all about deceiving you – and they and their self-centered doctrine HAS infiltrated and permeated ALL Religious Text to perpetuate the deceit!

My Father God and Savior – I love you beyond belief, and want to take this moment to thank you for loving me, *for loving us – all of us*, and for coming to this Earth to show us the way home.

It is important that we keep our eyes on The One True God and not let ourselves be deceived by the Miracles of Man: HAARP, Project Blue Beam, and others. *And others will come!*

Weird & Wonderful Tales

True Stories Just for You

By Lyn Murray

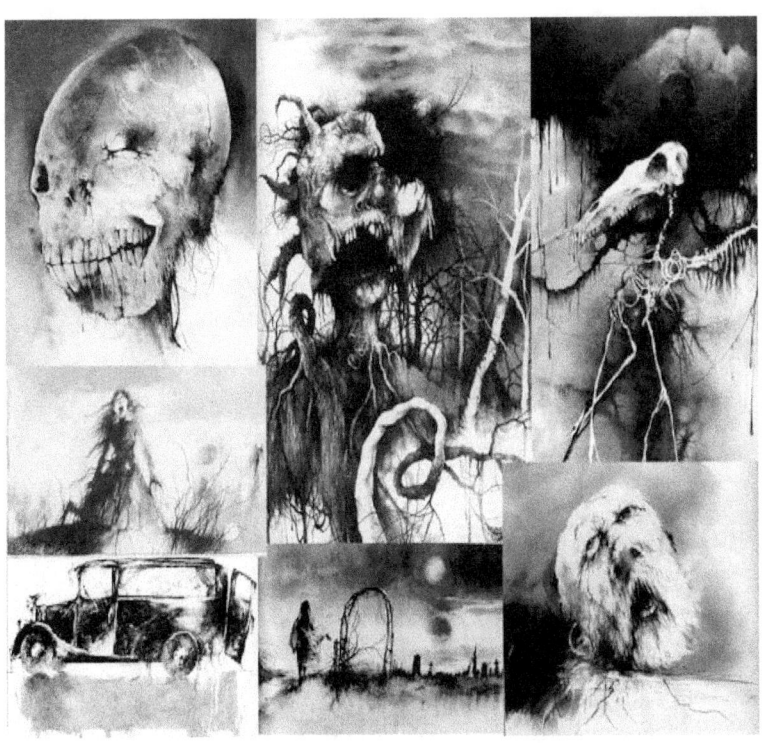

Hiccup in Time

This incident reportedly happened to a university student in Nova Scotia, Canada around 1991.

Apparently, he decided to take the express bus home one night to visit his parents for the weekend. He sat at the back of the bus and was alone - except for a family sitting behind the driver at the front. The bus ride was uneventful until it came time for him to depart.

As he looked out the window, he saw the tire factory up the hill. When the bus reached the top of the hill, he suddenly (and for some unknown reason) got a strange feeling that the family in front of the bus was laughing at him, that was accompanied by a strange "slip in time", where time repeated itself, and he (along with the bus and other occupants) suddenly found themselves about a mile back down the highway, and driving past the tire factory, *again!*

Of course, this scared him, and looking to the front of the bus, he noticed that the family, who had been talking before, was now *dead quiet.*

Standing up and approaching the front of the bus to get off, he told the driver what had happen. That's when the driver did the quirkiest thing: rather than looking surprised, he looked nervous, became sweaty, and stammered his response: "Things like that don't happen." he said, opening the doors, turning his eyes to the road ahead.

The young man said goodbye, yet the driver neither looked at him again nor spoke, but kept peering through the windshield into the dark road ahead as he exited the bus.

And although everything appeared to have returned to normal, when stepping off the bus — he had the strangest feeling that all this had happened before!

A Leap in Time

In the Fall of 1999, a young woman, and her boyfriend were on their way home to Crows Landing, California. There are only two roads into their town, from the north and from the south.

They approached from the south heading north (which is a four-mile, curvy two-lane road). It was a Friday evening around 6:00 p.m. and she was driving, and remarked how weird it was that they hadn't passed any cars – not any cars going in either direction!

Their hometown was a small town, but the two roads in and out were very well traveled roads. Even late at night, on any given night, you'd encounter cars coming and going.

Anyway, it was 6ish, when the car's dash, stereo lights, everything – suddenly went black, and the car died! At first they thought they'd run out of gas, but that couldn't be it because they'd stopped for gas and coffee before starting the last stretch for home.

Having no other choice, they coasted off to a dirt turnoff immediately ahead. Her boyfriend asked what happened? She replied that she had no idea, but that she felt drugged – like she was waking up from a long sleep.

When she put the car in park, her boyfriend jumped out, came around to the driver's side and got in while she scooted over to the passenger's side. Once seated, he cranked the engine, and it started just fine, and off they went.

As they pulled away (for some strange reason), she felt irritated, a little nervous, and absolutely sure that something really weird had just happened. That's when she glanced at the clock on the dash again and was shocked to find that an hour had passed. It was now a little after 7 PM; but according to her senses, the whole thing seemed to have lasted no more than one or two minutes.

So odd!

A lot of time has passed since this strange experience, but try as they have – they have NO recollection of what actually transpired, except for a few strange glowing spots on the back of the car, and oh, yes – dirt in the floorboard.

Phantom Church

A St. Louis couple (seeking solace in country life) was considering moving further from the city, and traveled to a hamlet called Rosebud, to look at a home on several picturesque acres.

They loved the house, and afterward decided to drive around to familiarize themselves with the neighborhood. It was a very rural area, and all the roads were gravel. They took one road after another until they were pleasantly lost in the beautiful countryside.

Deciding to turn around, they took a little gravel road that was in bad shape, and had to go real slow to keep from sliding; when they came upon a quaint abandoned church that sat off to the right. They couldn't take their eyes off it and crept by it really slowly looking at every detail.

The little church, now abandon, sat there all alone, its windows were gone, as was the paint on its wood exterior. The bell tower was still intact, but there was no bell. So dilapidated was the structure that they could look right through the front windows to those in the back. It was beautiful, kind of creepy, but beautiful all the same: the old pews were still there – sitting silently as if waiting for occupants to fill them. Sort of odd, as beautiful as they appeared to be, you would have thought they'd been stolen long ago.

As they slowly drove past, neither of them said a word, and just drove on in silence. They couldn't help thinking what a crime it was to let such a neat old structure fall into ruin. Then the husband said the strangest thing, "What do you think they were doing in there?" His wife didn't understand, and asked, "What who was doing where?" And he replied, "All those people."

Well, the gist of the story is that the wife saw no one – while her husband saw a church full of people sitting in the pews with their heads bowed down. After a couple of puzzling minutes trying to understand things, they turned around to have a second look . . . and this is where things get really strange!

No church!

At first, they thought they'd driven too far or gone down the wrong road, but when they recognized an oddly shaped tree they'd seen earlier, they knew they were going in the right direction. Stopping the car, they got out and walked around, but there was no church.

Suddenly and excitedly the husband called out for his wife to hurry and join him, and in reaching him, she noticed that he was as pale as snow, and pointing to the ground with a strange expression on his face.

She became a little pale too, when she realized that he was pointing toward what was an old pier and beam foundation that had supported a building about the size of the little church they had seen – *that was no longer there.*

The Conversation

Around 2:00 PM, a young woman's mother-in-law asked if she was going to pick up her six-year-old, first-grader at the bus stop or did she want her to? She remarked that the baby was still asleep, and asked her to go ahead if she didn't mind.

Of course, she didn't, and as she was walking out the front door, the young mother went to the back part of the house to check on the baby. That's when she heard her mother-in-law having a conversation with her mother as if she had just arrived and was letting her know what was going on.

She heard her mother's voice as clear as day and figured that any minute she would make her way through the house to say hello.

But to the young mother's surprise, that did not happen. Finding the baby safe and sound, she decided to go say hello to her, but in looking for her – she was nowhere to be found. Puzzled, she decided that she must have gone to the bus stop with her mother-in-law, and waiting for them to return, she thought no more of it. *Until . . .*

When her mother-in-law returned, she arrived only with the child . . . the young woman's mother was nowhere in sight. So, she asked, "Where's mom?"

"She must be at home. We visited most all day yesterday. Wish you could have been here. Did your ears burn? We talked about you the whole time." she said smiling.

Startled, the young woman turned white. "You weren't just talking to mom as you were leaving for the bus stop?"

"No dear. Why?"

The young mother couldn't speak and just sat down on the sofa.

What did it mean?

They had talked about her all day yesterday, so had there been a slip in time, a hiccup in time that allowed her to hear a past conversation?

We will never know!

Now You See It – Now You Don't!

One day while driving on I-40 in Tennessee, I kept noticing a couple of cars that seemed to be following me. Not that I was disturbed by this, oftentimes when traveling, like-minded folks travel in the same direction, and even frequent the same places for gas, lodging, and food. So, no, I wasn't bothered by this. What bothered me was the fact that these two cars seemed to pass me without me slowing down or stopping, only for me to notice a few miles up the road that they were mysteriously ahead of me.

This went on for hours.

Impossible!

When pulling into the gas station I couldn't help watching the occupants to see if anything looked out of the ordinary. It didn't, so I went about my business, filled up with gas, and went inside to grab a sandwich before continuing on my way. When I got back to my car, I noticed that my strange highway companions had already gone but in pulling back onto the highway, I noticed them pulling into the gas station behind me!

Impossible!

Now I'm freaked, and turned around to make sure that I was seeing things correctly; and sure enough – it was them: same cars (right down to the Garfield cat stuck to the side window of the van, and the pretty blond woman driving the truck).

When I pulled up beside them, the blond woman turned to me and smiled. Not a friendly sort of smile, a smile like "the cat that ate the cream" – smile.

When she turned to face me, I got the hell out of there and never looked back! I took another exit off the interstate and never saw them again.

Thank god! But, what was that? Who were they? *I don't think I want to know!*

The odd thing is – I still look for them in my rear view mirror. Odder still . . . is the irrepressible sense that – *I'll see them!*

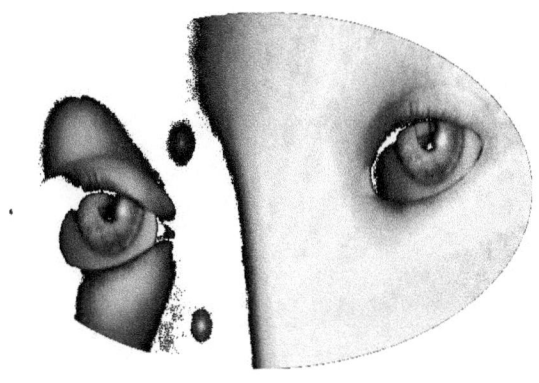

Nightfall's Day

Title Story

By Lyn Murray

October 6th, 10 AM

We set out early that Saturday morning. Kate and I, off on another day-trip to the country that would take an extended turn. We love day-trips. With our busy work schedules (corporate attorney battles for me, while Kate follows her dream of architectural design) day-trips were all we could manage.

Kate packed up the picnic basket the night before and left it sitting by the door so she wouldn't forget it like last time, and I packed up the car. All we had to do when the alarm sounded was shower, dress, and head out on another adventure. Which is exactly what we did.

On our way to the car, we encountered Mike Campbell in the parking garage of the high-rise apartment complex we all call home. Town Oaks – home sweet home, with an even sweeter price tag. We waved good morning and exchanged a few words across the concrete expanse of the parking garage, our voices echoing against the packed enclosure.

And we were off, the sound of our closing doors securing our fate.

We always carry a map and a flashlight or two, with plenty of batteries. GPS (Global Positioning System) is – despite what they tell you, unreliable, especially on back-roads travel. And regardless of what anyone might say, you always want to carry a flashlight because you never know when you'll get caught in a light-less situation. Like the time Kate decided we should explore an old cave on the outskirts of town, and our batteries went dead in the *one and only* flashlight we carried. What fun that was trying to find our way back to the surface with no light. And then there was the thing with the bats (that we could only hear because we had no light) – we still have nightmares, but I'll save the bat story for another time. All I can say is – never again! No, this time we were prepared for any and everything.

Or – so we thought.

But how do you prepare for the end when you don't even see it coming?

"Okay, darling, where to? You still set on the big city of Marveil?"

"You bet!" Kate answered back with her usual enthusiasm. "Take the drive down by the coastline. Like we planned. It should be beautiful this time of year. Everything is beautiful in the fall – don't you think, Peter?"

"Yes, I love fall colors."

Marveil is an artsy little community just down the coastline from Hampshire Heights – where we live and work. We'd heard so many good things about it from our friends; but until today, we had never quite been able to get there. Today that would change. Marveil here we come. We'd even made arrangements to stay over at a local bed-and-breakfast that Steve Winchell and his wife Sarah highly recommended.

Since we'd didn't take a honeymoon after marrying three years ago, (and with work, work, and more work being all we knew) and day-trips being our only entertainment – we decided, why not! Why not make this a trip to remember?

I recall how Kate kept looking at the bed-and-breakfast brochure Sarah had given her, remarking (sometimes to herself and sometimes to me) how lovely and charming it appeared – nestled away in the foothills of Cambridge Mountains, gently surrounded by a lovely grove of trees near Barstow River. A lovely river that seemed forgotten by time, with pristine waters that meandered through scenic mountain foothills and floral lined banks that would be ablaze with autumn's kiss this time of year.

We would river-raft there tomorrow.

The drive down the coastline was breathtaking. We had to stop several times just to drink in the view. The impatient ocean air was fresh and cleansing. About halfway (two hours into a four-hour drive) we decided to pull over for lunch and left the highway for one of those state roadside parking areas. Slim pickings for privacy, but the view was nice and we'd have abundant privacy once we reached Marveil.

By the time we parked, put the top up on my new sexy girl (my black Jag XK – my baby) and secured a picnic table, ominous looking clouds were forming over the ocean to the west. Wouldn't you know, the very direction we were headed. Strange clouds that seemed alive. Shelf like – with a mysterious, luminescent opacity to them.

Sitting down we faced the ocean to drink in the breathtaking view – *and to keep an eye on those clouds.*

Kate asked what I thought of them. "You ever seen clouds like that?"

I nonchalantly answered, "Can't say that I have," and took a bite of my chicken leg. "Dood, chicken, Hun. You've outdone yourself." I motioned for her to hand me another, asking, "Got any breasts in there?"

Kate grinned, "You know I do. Just enough for me, but I'll give you a bite," she paused to emphasize, "a *bite*."

"What is it with you and chicken breasts?," I asked. "I like them too, you know. A chicken comes with two. There are two of us," I said, motioning between us with my finger. "Why don't I ever get one, and why are there always six legs?"

"There are always six legs because I buy an extra pack of chicken legs, and they come four to a pack, that's why."

I understood that, what I didn't understand was the answer to the question I was about to ask for the hundredth time, "Why can't you buy an extra pack of breasts?"

"I thought you liked legs?"

And there it was – avoidance. Kate was a master at changing the subject.

"I do," I nodded.

"Then what are you complaining about?"

"I would just like a chicken breast once in a while – for Pete's sake (no pun intended) you don't own the sole rights to chicken breasts *ya know*."

"Yes, I do. Now hush and eat your legs. Here," she said, pushing the potato salad at me, "Get some before I eat it all and you have that to gripe about too."

I gave here my best growl, "Hand me the pepper – and salt!"

Kate passed the condiments, smiled and leaned in and kissed me on the cheek, and that's all it took. I melted like butter on a hot Summer day. Always do. Pushover. That's me.

Shifting her attention back to the gathering storm, Kate urged me to look at those clouds. "I mean it Pete, look. What kind of clouds are those?"

With a mouth full of chicken and potato salad, I grunted, "I don't know, but I don't think I want to get caught up in whatever they're full of. Let's finish up and get out of here, we can eat later. We need to get away from the coastline. We'll try and find a back-road to Marveil. I think it would be safer."

"You're probably right," Kate said, hurriedly gathering up food scraps for the trash; as I grabbed the picnic basket and headed for the car, yelling back, "Come on baby – I don't like the look of those clouds. Let's get out of here!"

"Coming."

But when I opened the trunk, the picnic basket was already inside. What the hell? A little taken aback I called out, "Kate?" But in glancing over to where she was standing, just behind the picnic table and a little to the right near the trashcan – I noticed the damnedest thing – a picnic basket, *our picnic basket*, packed and sitting on the table.

I'm thinking, shit! . . . (actually, more like – what the fuck's going on) as I looked back in the trunk to find no basket. But when I looked back at the table there was no basket there either, and then I realized that I was still holding it – just as I remembered. Startled, I stepped back, dropping the basket.

"Shit!"

By this time Kate had reached the car, and seeing that I was rattled, she asked, "Peter? Are you okay? Close the trunk. Let's go."

I'm thinking – I'd love too, and trying to collect myself I reached down to pick up the basket – but it was gone.

Kate walked around me and started closing the trunk lid, when I stopped her. And there it was. The picnic basket – inside the trunk. I stepped back.

"Peter. . ." Kate closed the trunk lid and took my hand. "You okay sweetie?"

I looked at her – amazed that she was oblivious to everything that had just transpired. Not knowing what to say I shook my head and said (as calmly as I could, not wanting to scare her), "Yeah, yeah, I'm fine. Got a headache coming on. I'm fine."

◇◇◇

Once in the car, Kate checked the map to see if she could find a nearby back-road to Marvail, and she did find one that was about a mile up the coast road, to the left of a dot on the map called Merchant.

Working our way through the winding roadside park drive, we pulled out and back onto the main highway.

◇◇◇

"What did you say?"

I glanced over at Kate who was shaking her head, no. "Nothing, I was just thinking that there aren't many cars."

"I could have sworn that's what you said."

"Nope, just thought it."

"If you say so." Again I'm thinking, what the hell?, but, I said, "You're right, there aren't many cars. They probably pulled off because of the storm. I would too, but in light of the storm surges we've had lately, I'm a little hesitant. I don't think we want to be this close to the ocean if a big wave hits, and those clouds look like they could produce some nasty waves. I think it best we get out of here. . . . Just hope we can make it to Merchant before it gets bad."

Crossroads aren't always clear – and unknowingly we were at one.

"Okay, if you think we should."

"I do. We'll be okay. Buckle up!"

"What did you say?"

"I said we'll be okay. Buckle up."

"Are you sure that's what you said? I could swear you asked me to check the map for a back-road."

"No, I asked you that before we pulled out of the rest stop. Remember?"

"Yeah. That's why I thought it strange when you asked me again."

"But I didn't."

"Okay, my mind must be wandering."

I'm freaking out inside. We were experiencing far too many déjà vu's to chalk it up to mere happenstance. Something was wrong. Something I didn't have a name for. Gotta hold it together. Mustn't let Kate know how freaked out I am. Hold it together dammit! Just hold it together.

As I pulled back onto the highway it started to rain, not a sprinkle here and there – pouring down rain. By the time we reached Merchant (although it was only 2 PM) it was as dark as any 9 PM I'd ever spent during winter months down south. Every car we met on the highway (that we could see) had their headlights and dimmers on. *It was that dark.* And what traffic there was crawled along. Between the wind and rain vision was practically zero, (and whishing more than a few times that we'd driven the truck) we decided to wait it out and pulled beside a country store at the Merchant cut off where we parked.

On the way through the small parking lot that led to the side of the building, we remembered seeing a big sign near the drive's entrance that read 'Fresh Hot Coffee', and in desperate need of a caffeine fix we debated trying to make a run for it – when reason lost to need, with us exiting the car as we made a mad dash for the door.

The small store (that didn't look much larger than our kitchen) had obviously survived previous weather surges, and with the storm worsening we felt we'd be safer inside than outside – even though the building looked like it was held together with toothpicks. And once we reached the door we shuffled in like a couple of drowning rats that were seeking higher ground.

Once inside we were greeted with, "Get in here. Get in here." Instructions that were delivered by a squeaky little voice that sang out like Maw Kettle yelling for Paw and the kids.

"My goodness. Where'd you come from? Why didn't you pull over before the storm hit?"

Reminding me of my mother, a flashback to my younger days gave me pause – but alas . . . it was just another brother's mother. It's a mother thing – undoubtedly they all have the same programming.

Soaked to the bone and squishing inside my shoes, I answered as politely as I knew how, "Oh, you know, so close we hated to stop. Thought we could make it here before it got too bad. Guess we were wrong, huh." Chuckling, I helped Kate in and shut the door.

"No bother. I've got just the thing. You come here and sit by the stove and dry off while I get you a cup of coffee." And handing us a roll of paper towels, she asked, "You do drink coffee?"

I laughed, "Yes, ma'am. That's why we're here. That's quite a sign you've got out front. Hard to miss." And heading to the warmth of a potbelly stove, I added, "We'll take two cups of the hard stuff. None of that decaf . . ."

"Don't drink it, don't make it – it's un-American," she said grinning, her pudgy little face lit up with a smile full of dimples. "Well, you just sit right down and make yourselves at home. Cream and sugar?" The cute little gray-haired woman (as round as she was tall, which looked to be four feet) shouted over her shoulder as she hurried back to the front of the store.

Kate shouted back, "Two creams, one sugar for us both."

"I'll be right back," she said, disappearing around a rack in front of the coffee pot that sat atop the end of a rustic wood counter that looked as old as she did. "You folks from around here?"

"Not really," Kate said, shouting over the storm. Had to shout – the building's tin roof amplified every drop of rain and wind gust to the point of drowning out everything else. "Just passing through. On our way to Marveil."

I was freezing and warmed my hands by the stove. (A Dearborn heater. I hadn't seen a Dearborn heater since the last Christmas at Grandma's house in '92. "Wow . . . didn't know they still made these." I yelled to the rafters, thinking it's probably as old as she is. Just well kept.

"Oh, my, Marveil is such a cute place. Been there once myself. Long time ago – when Papa was alive. Don't get out much anymore. Not since Papa died . . . two years ago last spring."

"Sorry to hear that," Kate said.

"Me too, dear. Me too. Life goes on." The squatty little woman said, walking back with two big cups of steaming hot coffee, which she quickly handed us before disappearing through a small door in the back, to just as quickly reappear with two plates of biscuits.

"Here you go. Made just a few minutes ago. Still warm. Made from scratch mind you. Not canned. Hate those things. What a waste of dough."

And there was that smile again. What a charmer.

"Butter's bought, but it's Blue Bell – so it's almost like homemade. Good and sweet." she said, handing us each a plate. "You eat up. It'll be good with your coffee."

And it was. We especially liked the homemade plum jam she dipped out onto our plates.

◇◇◇

As we sat there eating biscuits and drinking the best cup of coffee we'd had in a long time (while warming by that old Dearborn heater) the storm raged outside, casting a late afternoon spell over the quaint little shop that crept up our spines with an eerie foreboding.

Our hostess pulled up a chair alongside ours and introduced herself. "I've got the worst manners . . . Martha. Martha Danvers. I own this place. Lived in the back ever since Papa died. Just too much trying to keep up a house and business too."

Kate spoke up, "I'm Kate. Kate Westfield. This is my husband, Peter. Peter Westfield." *I could only smile and nod because my mouth was crammed full of biscuits and jam.*

"Nice to meet you," she said, taking a deep breath. "Whew – it's nice to sit a minute. Been on my feet all morning. You don't realize how tired you are 'till you sit down."

Swallowing my last bite of biscuit, I looked toward the front of the store and commented, "That's some storm. Hope it clears up soon, we've got reservations at Bethany's Bed and Breakfast in Marveil. Need to check in before 6 PM or we'll lose it."

"Oh . . . I'm sure they'll hang onto it for you – in light of the storm and all."

"I tried calling but couldn't get a signal. Your phone work?"

"Nope. Went out a few seconds before you arrived. Usually does with storms. I'll check again in a few minutes. Never know, they might get it working again before nightfall."

"Okay. That'd be great. So what can you tell us about Marveil?"

"Oh, it's cute. Really beautiful! Reminds you of a Norman Rockwell painting. Sits in the foothills of Cambridge Mountains. Lovely place. Lots of artist folks live there."

Kate smiled. "So we heard. Some friends stayed there a year or so ago – and they just raved about it."

"You'll love it. Papa and I sure did." She paused. "Lord, that must have been the late 60's," she said, gently shaking her head while trying to reconcile all the years that had so quickly slipped by.

Old couldn't have been any cuter.

We must have sat there an hour or more trying to outwait the storm, and kept ourselves busy sipping coffee and eating biscuits, listening to our hostess, and sharing memories. When the storm broke sending sunlight through the storefront windows we knew it was time to go; and saying our goodbyes to the kind, dimpled face we'd only just come to know, a strange sensation tugged our heartstrings and gave us pause. I couldn't help thinking how much our hostess reminded me of my grandmother. I hated to leave – we both hated to go. It was one of those "ships passing in the night" moments that you remember forever.

What a sweetie.

Once we were back in the car and buckled up, we plotted our back-roads approach to Marveil. I can't explain why we didn't just get back on the main road. It would have been much simpler, but we didn't.

Regrets are like hindsight - you never see them coming until they land squarely in your nightmares.

"At least the road is paved," Kate said. "Weren't those biscuits good?"

"Sure were. They hit the spot."

"Sure did. I'm going to miss her. Such a sweet little thing."

"Well – maybe we can come back for a visit. I'd like that. I think she would too."

"I think she would," Kate said rattling the map, "this doesn't look familiar," she said turning it upside down. "I think we're going the wrong way."

"Can't be, the compass shows due West."

"I don't care what it shows. This feels wrong. Don't you feel it?"

"Feel what?"

"That this is wrong."

If I'd learned anything after three years of marriage, it was to stop and listen to your wife when she sounded worried. So I stopped the car. "What is it?" *She didn't answer.* "Let me see the map," I said, pulling it from her hands.

Kate absentmindedly handed the map over without looking in my direction because her gaze was captured by something outside the passenger's side car window.

"Pete. . ."

"Kate?"

"Pete. . . What is that?"

Buried in the map, I asked without looking up, "What is what?"

"That!"

Something about the tone of her voice said – *look up.*

"Peter. What is that?"

It's a small car and the very presence of her in the passenger's seat kept me from being able to see out her window, so I leaned forward a little trying to see past her.

"Do you see it?"

"No, baby, I can't see a thing. Sit back a little, you're blocking my view."

She moved like molasses on a cold day. "Honey, sit back so I can see what you're talking about."

"Peter. What the hell is that?"

"I don't know. Sit back so I can see what you're talking about."

She suddenly turned to me with panic in her eyes and a face that was as white as a sheeted ghost.

"Peter. Drive! Just drive!"

"Kate?"

I wanted to see what she was talking about and struggled to see past her and out the window. "Kate . . ." I pushed her back in the seat. "Kate . . ."

And there it was, the strangest cloud I'd ever seen. Almost like, no . . . exactly like the cloud we'd seen coming in over the ocean as we ate lunch at the roadside park. The cloud that brought that storm. That strange opalescent cloud. *But low now.* Ground level low. Just sitting there – alone, not part of a larger system of clouds. Just alone. Not moving, but moving. Folding in on itself. Reforming. Transforming.

"What the hell . . .?"

"Get us out of here, Peter! Now. Go, just go!"

I had a sense of urgency too, but as I tried to put the car in gear it wouldn't go. The car was still running, I had only pulled to the shoulder and put it in park, but now I couldn't get it to shift.

"*Peter!* For the love of God, get us out of here. It's moving our way. Get us out of here. Now Peter, now!'"

"I'm trying, Kate, the gearshift won't engage."

Kate screamed . . . *"Peter!"*

Suddenly the car was engulfed in that cloud. It was everywhere. All around us. Nearly pitch black except for a strange glowing luminescence. I grabbed Kate and pulled her close to me. Terrified she buried her face in my shoulder.

"Peter. Peter." *The tone in her voice resonating every fear I dared not speak.*

Everything went black.

———

As I pulled back onto the highway it started to rain, not a sprinkle here and there – pouring down rain. By the time we reached Merchant (although it was only 2 PM) it was as dark as any 9 PM I'd ever spent during winter months down south. Every car we met on the highway (that we could see) had their headlights and dimmers on. *It was that dark.* And what traffic there was crawled along. Between the wind and rain vision was practically zero, (and whishing more than a few times that we'd driven the truck) we decided to wait it out and pulled beside a country store at the Merchant cut off where we parked.

That's when it hit me.

"Kate? We've been here before."

Kate laughed. "What are you talking about? We've never been here," she said, turning her gaze outside.

"We just left here." *A gruesome chill gripped my spine.* "Kate, I'm serious. We just left here. Not an hour ago." I looked at my watch but an hour hadn't passed, so I asked Kate to check hers.

That chill crawling down my spine intensified.

"What's wrong with you Peter? Stop it."

"I'm serious. Please – just check your watch. Try to remember."

"Stop it, Peter, you're scaring me. What's wrong with you?" She took my hand. "Are you feeling all right? Come on, let's go inside and have a cup of coffee. We didn't have breakfast. *No coffee.* You just need a caffeine fix and you'll be right as rain. Come on," she said opening the door.

Maybe she was right. Maybe I was dreaming. Maybe it was one of those déjà vu moments. Sure, that had to be it. She had to be right. That's crazy, I'm thinking. Sure – she's right. What else could it be? But that chill that had once gripped my spine now clawed its way to the pit of my stomach. *I felt sick.*

Getting out of the car we covered our heads and tried to shield ourselves from the wet stuff and ran to the front door and stumbled in like a couple of drowned rats. We'd no sooner stepped in when we were greeted with a cheerful urging to come in out of the rain. "Oh my – get in here. Get in here." A squeaky little voice sang out like Maw Kettle yelling for Paw and the kids.

"My goodness. Where'd you come from? Why didn't you pull over before the storm hit?"

Mom? – I jokingly thought. Couldn't be . . . just another brother's mother. They must all have the same programming.

That's when I froze, *stopped dead in my tracks,* which must have been a more pronounced halt than I realized because Kate turned and looked at me with a dazed expression.

"Peter?! My god, what's wrong? Peter? You're white as a sheet! Peter?!"

That made two of us.

That was six months ago. I think it's been six months, it's hard to know with time as fucked up as it is? You see, I didn't start counting right away, too stunned – but around six months ago is when the realization of looping time began. I've settled down a bit since then. You adapt, if for no other reason than for the sake of

sanity – you adapt. If not totally, somewhat. But I'm not making any sense, am I . . .

◇◇◇

October 6th, 4 PM

"Well maybe we can come back for a visit. *I'd like that.* I think she would too."

"I think she would," Kate said, rattling the map, turning it over and upside down. "This doesn't look familiar. I think we're going the wrong way."

"Can't be, the compass shows due west."

"I don't care what it shows. This feels wrong. Don't you feel it?"

"Feel what?"

"That this is wrong . . . "

◇◇◇

October 6th, 4 P.M.

Yes – this is wrong; but how can I make her understand? She doesn't hear me now. In effect – she's not my Kate. She is but she isn't. *How the hell do I comprehend this?* She's not living the same day that I'm living over and over. She's trapped inside another loop in time – while I'm outside – trapped in another. We're sharing space only in that we're allocated to this site – but we're not connected any longer. She's talking to someone who is – but isn't me. Not the me that occupies the space that I'm cognizant of. But unlike me – she's not mindful of being trapped. How is that possible? *Oh God!*

Why is it that I am aware, that I can remember?

God . . . why didn't I major in science?

Searching for answers I remembered something I'd read about an electronic monitoring system that has people stirred up and bitching because they believe it's responsible for all the strange weather we're having around the world.

What's it called? Damn!

It's supposed to have been developed for the purpose of analyzing the ionosphere and investigating the potential for developing ionosphere enhancement technology, supposedly for radio communications and surveillance – but scientists are finding it's capabilities go way beyond that! *Dangerously beyond.*

HAARP! That's it. I wonder . . . ?

Every day the same day. Sometimes it begins in the same place, sometimes in another, but always the same day – and I don't know how to stop it. Maybe with that cloud. Maybe if we could avoid that cloud. Maybe that would reset time. Maybe?

Nightfall's day, nightfall never comes but it's always there. An endless day, a day without night. *Waiting.* I can't help thinking that if it came, if night came, maybe this nightmare would be over. *Would it?*

Time. Something happened to time. But what? Something in that cloud altered time, changed it somehow. Without the cloud, were we lost?

How many days has it been? How many repeated events? How many cups of coffee with the same old lady, in the same store, gathered around that Dearborn heater? How many?

Oh God - how *many* more?

October 6th, around 4 P.M.

"Peter. For the love of God, get us out of here. It's moving our way. Get us out of here. Now, Peter, now!'"

"I'm trying, Kate, the gearshift won't engage."

"Peter!"

Kate screamed.

October 6th, 4 P.M.

Something has changed. Is changing. It started a few days ago. A few days? It could be a few months of days for all I know. It happened gradually (not all at once) as if someone was trying to conceal the change. Someone or some "thing". But who, or what? And why?

The change was subtle but change there was, because every time we stopped on that back-road, every time, every day, as Kate checked the map after we pulled over, we were parked a little farther up the road than the day before. Maybe only a few feet (a few inches) but nevertheless farther up the road than the day – the time before.

How do I know? Because I started marking the landscape. Landscape is stationary, it doesn't move, it only changes location when the observer moves, so we were definitely moving – but to where? How? And why weren't we aware of it in real time? Was time trying to reset itself? Time as we know it? Was our slow progression toward an unknown destination part of the resetting process? Were we just looping, and not really moving at all? Was I the "me" that began this journey – or another "self" caught up in an endless stream of possible outcomes?

Dammit! HAARP! That has to be it! Oh My God – but then there's that SuperCollider in France. Cern! The Hadron SuperCollider. Damn – what if. . . ? They were supposed to fire that thing up the 23rd of September (2015). What if they did. What if. . . ? Damn them – fucking scientists – they did this! Just because you can . . . doesn't mean you should! *And these are the smart guys?!* The ones running the world? No wonder we're so fucking screwed up.

I shouted at the top of my lungs, "You've really done it this time, haven't you?! What if we can't come back, what if you've undone time, unraveled it somehow? How the hell do we get back? You don't know do you? You didn't think that far ahead did you?! *Bastards!*"

◇◇◇

I've had a lot of time to think about it, probably way more time than I realize. Could this road have something to do with the anomaly we're trapped in? If we'd never come up this road, would the time distortion have occurred? Would we be trapped in it? Is everyone trapped OR are we the only ones? If I could get off this fucking road – out of the area . . . Something about this area . . . where is HAARP situated? Do they have more than one array? Dammit!

I thought back to the roadside park where we ate lunch, something was starting then – because we kept thinking we were talking, saying things, seeing things we weren't, or weren't aware of on a conscious level, but those anomalies were hiccups in time. That cloud over the ocean. *The precursor.* But how did that affect everything, everyone else, or was that cloud the actual anomaly that now engulfed us? We weren't in this alone, couldn't be, others had to be involved. But if we're having trouble communicating with each other, how would we communicate with them – even if we encountered them?

Nevertheless something's happening. Something's different. Today, when we reached the Merchant cutoff to Marville,

Mrs. Danvers didn't serve biscuits with our coffee.

In fact – there was no Mrs. Danvers. And that's not all . . . there was no store, either.

What the hell was going on? Temporal shifts? They're going to kill . . . kill us all! While they're playing around with the atmosphere, Mother Nature's taking defensive action. Maybe everyone's dead. Maybe we're all dead, and just don't know it. Trapped somewhere in time – *by* time? God!

What will tomorrow bring? *Tomorrow?* Today!

I'm afraid to ask, afraid to think about it. Why? Because Kate's starting to fade, or is it I? She doesn't seem to notice. She acts as if we're talking, but I don't think she can hear me anymore. Not me, not the me I am – *was*; but if not me – then who? Another version of me – that's trapped in another dimension? An alternate reality? God! Is that even possible?

My God!

◇◇◇

October 6th begins again, farther up the road.

Where the hell were we? Was this the same road? It looked different, although I hadn't driven anywhere.

He called out to his wife, "Kate . . .?"

There was no reply.

He looked toward the passenger's seat...

Temporal Displacement

Temporal Displacement

What's the problem?!

The purpose of this study is to explore the injurious effects of temporal displacement upon individuals (and society) who've experienced it and to suggest viable methods of treating such individuals.

A General Survey was used to examine a conceptual framework that suggests the repercussions of hypothetical temporal displacement, by exploring actual cases (largely over two centuries), and establishing a model that was activated by problem and self-transcendent awareness, triggered by the individuals' experiences.

First, a little background:

As a society, we have long fantaszied time travel, for something like 700 years. Writers from as early as the nineteenth century have explored the possibilities of temporal shifts. In some of the tales written, the mode through which characters passed through time was hazily depicted.

One prominent American satirical author of the 1800's, Samuel Clemens (nom de plume: Mark Twain) moved his character, Hank

Morgan, through time from 1889 Britain (i.e. Europe's Northern Island) to the year 528, when the thought to be fictional King Arthur reigned over the Knights of the Round Table. The hero of Clemens' story, Morgan is transported by means of a dream, as was the usual custom for the era. Approaching temporal shifts in this manner, alleviated the writers' need to explain the yet undiscovered conditions under which temporal displacement occurs. Convenient, but not totally satisfying.

While living in these ancient times, Morgan introduces the telephone wire. He also seeks to change the ineptitudes of the feudal system by introducing schools, mechanical devices, and the democratic values of his own time. The trick of returning him to his own time is solved by Merlin, a famous wizard of King Arthur's day, in which the spell he casts upon Morgan causes him to go into a deep sleep until he awakens in his own time.

Another prominent example of fanciful conjecture about time travel was H. G. Wells' *The Time Machine* (*one of my personal favorites!*). The story was written in 1895 and popularized to an even greater extent through the 1960's film starring Rod

Taylor as the audacious time traveler. The image of a device that moved a person through time fueled the imagination of writers and pseudo-scientists for years to come.

Although Wells' story is one of fiction and not based upon any real science, he did manage to project a degree of insight as to the possible effects temporal displacement may have on an individual's psyche.

In fact, stories of "time travel" became so prevalent during the late twentieth and twenty-first centuries that many people believed it to be an inevitability, as tales of time warps, storms, leaps, slippages and tears became fodder for the imagination. Reality, of course, makes science fiction irrelevant – as science fiction makes science irrelevant. Each *stand alone* in their field.

Though most peoples' speculations about temporal displacement are profoundly flawed, the fascination with time travel inadvertently provokes people (in particular science fiction writers who stand to gain financially from their sometimes enigmatic presumptions) to consider some possible negative consequence of temporal displacement. However, the human elements of a number of such conjectures come close to actual issues and need to be taken more seriously.

The world we live in today is vastly different from that of those who penned such imaginings. Today we have real aspirations and expectations of time travel – with real science to back them up. As space travel becomes more prevalent, so does the dream of time travel.

There has been vocal and ongoing controversy among scientists, psychologists, politicians, recreational facility developers, social geographers, historians, and ethicists for more than 100 years as to whether time travel – at least, as we understand it – should be developed and fostered. Fortunately, efforts to win support for further investigation of the potentialities of temporal displacement among a substantial number of the aforementioned minor factions, continue to fail across the board on a global level. Most of society's citizens rightly recognize the clamor of many of these groups as a pretext for exploitation. For those wishing to push through the research, problematic heightened awareness concerning this issue (after the first "presumed true" documented cases of actual occurrences of temporal displacement recognized several times during the mid-21ˢᵗ century with the advent of video cameras and cell phones) have highlighted the motivations behind ego-centric arguments in favor of such research.

This author believes that the ability of time travel is already known (and practiced) and held in strict confidence by world powers, and that the only ones on planet Earth who don't know about and use this as a means to get from one place i.e. one time to another – are the *grunts* of society – which make up the majority of the population.

For the universal basic temporal displacement theory ("time travel"), I offer the following explanation:

Temporal displacement, as far as modern man understands it, *can only be accomplished* by an individual or group's contact with a *wormhole*:

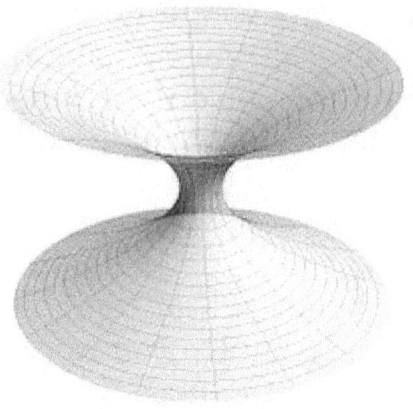

$$G_{\mu\nu} + \Lambda g_{\mu\nu} = \frac{8\pi G}{c^4} T_{\mu\nu}$$

"Einstein-Rosen Bridge":
(Based on Einstein's General Theory of Relativity.)

"A **wormhole** is a hypothetical topological feature that would fundamentally be a shortcut through space-time. A wormhole is much like a tunnel with two ends, each in separate points in space-time.

For a simplified notion of a wormhole, visualize space as a two-dimensional (2D) surface. In this case, a wormhole can be pictured as a hole in that surface that leads into a 3D tube (the inside surface of a cylinder). This tube then re-emerges at another location on the 2D surface with a similar hole as the entrance. An actual wormhole would be analogous to this, but with the spatial dimensions raised by one. For example, instead of circular holes on a 2D plane, a real wormhole's mouth could be spheres in 3D space.

Researchers have no observational evidence for wormholes, but the equations of the theory of general relativity have valid solutions that contain wormholes. Because of its robust theoretical strength, a wormhole is one of the great physics metaphors for teaching general relativity.

The first type of wormhole solution discovered was the Schwarzschild wormhole, which would be present in the Schwarzschild metric describing an eternal black hole, but it was found that it would collapse too quickly for anything to cross from one end to the other. Wormholes that could be crossed in both directions, known as *traversable* wormholes, would only be possible if exotic matter with negative energy density could be used to stabilize them.

The Casimir effect shows that quantum field theory allows the energy density in certain regions of space to be negative relative to the ordinary vacuum energy, and it has been shown theoretically that quantum field theory allows states where energy can be *arbitrarily negative* at a given point. Many physicists, such as Stephen Hawking, Kip Thorne, and others, therefore, argue that such effects might make it possible to stabilize a traversable wormhole.

Physicists have not found any natural process that would be predicted to form a wormhole naturally in the context of general relativity, although the quantum foam hypothesis is sometimes used to suggest that tiny wormholes might appear and disappear spontaneously at the Planck scale, and stable versions of such wormholes have been suggested as dark matter candidates. It has also been proposed that, if a tiny wormhole held open by a negative-mass cosmic string had appeared around the time of the Big Bang, it could have been inflated to macroscopic size by cosmic inflation.

The American theoretical physicist, John Archibald Wheeler coined the term *wormhole* in 1957; the German mathematician HermannWeyl, however, had proposed the wormhole theory in 1921, in connection with mass analysis of electromagnetic field energy.

This analysis forces one to consider situations . . . where there is a net flux of lines of force, through what topologists would call "a handle" of the multiply-connected space, and what physicists might perhaps be excused for more vividly terming a "wormhole"."

—John Wheeler in Annals of Physics

While entertaining, a Connecticut Yankee in King Arthur's Court (a novel by Mark Twain, 1889) had less to account for than a present-day space traveler would have to explain about a modern space vessel, even to a much more "enlightened" man from a more technologically advanced time. Today – we need more information to call it *informed* entertainment!

Current and future time travelers (both those who travel willingly, and those who are seemingly and mysteriously transported through time – by accidentally encountering a "worm hole", could be transported to totally unfamiliar and unrelated space and time – space and time that may or may not be related to Earthly destinations. *What a concept!*

As yet (or at least to the *grunts* of society) apparently there is no way to predict the time and final destination of an individual's travel. However, as stated – this author believes the governing factions of this world and beyond DO have timeline capabilities, and can project themselves when and where they choose! Whereas those of us who accidentally get caught up in one of the wormhole projectiles – could be flung from the surface of the Earth to anywhere in this universe and beyond – which would (most probably) be a one way trip ending in a death sentence. However, depending upon the worm hole's location and stability – it might just spit you back out in the vicinity of origin.

No one can ascertain how long the period may be between the transference of a being from one time and place to the approximate point of origin (which no doubt will result in many lost lives due to the process of learning). Whether or not their journey is instantaneous or delayed, which brings about a whole new set of tantalizingly torrid questions to perplex generations to come.

In consideration: as a whole, time travel is both an intriguing and frightening concept in this modern age. The issues of altering history, introducing technology, and ideology before its time – alone, are enough to keep you up nights wondering if your stock's sudden decline had something to do with an errant time traveler's competitive stocks rapid rise. And don't think it's not possible – it's very likely probable. *"He who controls "time" controls everything within it!" – Lyn Murray.* It may have already happened:

Wall Street Time Traveler

"Bizarre - 'Time-Traveler' Busted For Insider Trading

From The Weekly World News
By Chad Kultgen

3-28-03

(Note from Lyn: although you probably won't find evidence of this story anywhere but here . . . it was aired by mainstream news stations (I saw it!) – before disappearing forever as part of the

"disinformation act" that is perpetrated against the American people! If authorities had nothing to fear – why did this story disappear, only to remain on a "conspiracy theorist" site in an attempt to debunk it?!)

Andrew Carlssin being led away by the FBI after his arrest for Insider Trading. This is a "still" shot of the actual video of his arrest – which was aired on FOX and CNN, and others!

▼

Reminder: some of the following "actual files" (which were collected just for you and left as original as possible) may download oddly, depending upon which eBook reader you use – however, they translated beautifully to print, so you may want to pick up and keep a hardcopy of this book (while sharing the eBook) if you are a lover and collector of weird historical facts that have a *tendency* to disappear over time.

Above is reported to be an archival picture of citizens attending the reopening of the South Fork Bridge in the early 1940's, Gold Bridge, B.C.

Everyone always focuses on the man with the out of place/out of time sunglasses. Look at the highlighted picture of the man in back. I don't know about you – but that looks an awful lot like our Time Traveling Wall Street Insider Trading friend . . . *Andrew Carlssin!*

▼

Now for the story:

"NEW YORK -- Federal investigators have arrested an enigmatic Wall Street wiz on insider-trading charges -- and incredibly, he claims to be a time-traveler from the year 2256!

Sources at the Security and Exchange Commission confirm that 44-year-old Andrew Carlssin offered the bizarre explanation for his uncanny success in the stock market after being led off in handcuffs on January 28.

"We don't believe this guy's story -- he's either a lunatic or a pathological liar," says a SEC insider. "But the fact is, with an initial investment of only $800, in two weeks' time he had a portfolio valued at over $350 million. Every trade he made capitalized on unexpected business developments, which simply can't be pure luck." *Not unless you live in fantasy land!*

"The only way he could pull it off is with illegal inside information. He's going to sit in a jail cell on Rikers Island until he agrees to give up his sources."

The past year of nose-diving stock prices has left most investors crying in their beer. So when Carlssin made a flurry of 126 high-risk trades and came out the winner every time, it raised the eyebrows of Wall Street watchdogs.

Get it? If this is true (and, remember, I saw the news report with my own eyes — I'm not just writing about things I haven't personally witnessed) — had he "not been" outside the Rothschild Loop . . . he would have walked!

"If a company's stock rose due to a merger or technological breakthrough that was supposed to be secret, Mr. Carlssin somehow knew about it in advance," says the SEC source close to the hush-hush, ongoing investigation.

When investigators hauled Carlssin in for questioning, they got more than they bargained for: a mind-boggling four-hour confession.

Carlssin declared that he had traveled back in time from over 200 years in the future, when it is common knowledge that our era experienced one of the worst stock plunges in history. Yet anyone armed with knowledge of the handful of stocks destined to go through the roof could make a fortune.

"It was just too tempting to resist," Carlssin allegedly said, in his videotaped confession. "I had planned to make it look natural; you know, lose a little here and there, so it doesn't look too perfect. But I just got caught up in the moment."

In a bid for leniency, Carlssin has reportedly offered to divulge "historical facts" such as the whereabouts of Osama Bin Laden and a cure for AIDS.

All he wants is to be allowed to return to the future in his "time craft."

However, he refuses to reveal the location of the machine or discuss how it works, supposedly out of fear the technology could "fall into the wrong hands." *Smart AND rich – my kinda guy!*

Officials are quite confident the "time-traveler's" claims are bogus. Yet the SEC source admits, "No one can find any record of any Andrew Carlssin existing anywhere before December 2002." (*Note from Lyn: it isn't clear whether or not he was carrying ID – however, even in this day and age – that can be faked.*)

Weekly World News will continue to follow this story as it unfolds. Keep watching for further developments."

It is this author's opinion that there is more truth than fiction here (*since I don't believe in coincidences, and if knowing I could benefit from 126 specific stocks on specific days is coincidental – I'll kiss your posterior downtown in broad open daylight!*) and since authorities cannot find any evidence of this man's existence prior to his arrest! *That cinches it for me.*

Also, since breaking his probation – he is

nowhere to be found!

Perhaps, this is why there is such a financial/wealth distribution gap between the have's and the have-nots! This issue has been discussed ad nauseam throughout the past seven centuries, at least! This author wishes to reiterate that the purpose of this passage is to confront and offer solutions that address the psychological effects of temporal displacement/time travel upon society and the individual psyche.

In conclusion:

Although temporal displacement is rare and unpredictable (for the majority of us) its effects may be both long-term as well as damaging to an individual's psychological frame (not to mention biological side effects). Since the event itself cannot reasonably be avoided, it is imperative that strategies is developed to deal with the after-effects of temporal displacement.

Three specific areas that must be addressed by a counselor are: 1) a subject's relationships with others in an alternate time line, should this be his/her experience;

2) the problem of confusion, i.e. helping the victim of temporal displacement discern what is real and what is not; 3) helping to alleviate their guilt for actions that they are not accountable for.

In order to meet this need, I propose a new course of study be developed in the Counseling Department of at least one major university. Since the problem of temporal displacement applies to so few people (at present, at least) only a small number of competent counselors would be needed to follow this course of study. I, therefore, propose that candidates accepted into the program be an elite group of perhaps five individuals. The question may be raised as to whether there truly is a need for such counselors and whether it is expedient to take from the best in order to meet the needs of so few. This author answers with a resounding – *yes! Of course, doing so would be admitting that Time Travel existed!*

Since we are still unaware of the consequences of people being hurtled into another time and space, it is imperative that we take adequate precautions when considering the issue of temporal displacement. And this author dares to submit that the only substantive insights to be gained will be by learning as much as we can from people who have experienced the phenomenon.

I leave the issue of how to handle people who are displaced (to our time from another) to others who wish to tackle this sticky question. However, given the accounts recorded here, it seems clear that we have no choice (as a society) than to prepare ourselves for additional occurrences of temporal displacement, as the intentional displacements could result in a "domino effect" and begin to open up involuntary wormholes that begin to distort time and space for all of us. If we act now, perhaps we will help a few. On a grander scale – we could aid the entire human race. Only time will tell.

Notes:

Several significant questions have not yet been adequately addressed concerning the possibility of temporal displacement, and our ability to spot those traveling through time. Unless – like our Wall Street friend – they screw up, and their greed gets them caught!

Question: who is monitoring the heads of

these large multibillion-dollar corporations – *for example?* Who is to say – the heads of these corporations aren't in fact Time Travelers foreseeing our needs and capitalizing on our unilateral decisions to act upon them? Which isn't to suggest that they have anything but their own interests at heart – usually at our expense . . . like Monsanto, Bayer, and the rest.

Bend over and pass the vaseline – please!

Still not convinced?

Then let's examine

Andrew Basiago:

Time Traveling Child

The US time travel cover-up is blown wide open with Andrew Basiago's story revealing his childhood involvement with the U.S. Government time travel experiment back in the 1960's

Andrew D. Basiago is today a successful lawyer in Washington, writer, Mars researcher, and planetary whistle-blower, as he claims to bring the truth about Time Travel to the world as he recounts his childhood experiences in DARPA's Project Pegasus.

The photo above is supposed to be Basiago when he traveled back in time to witness President Lincoln's Gettysburg Address.

In Basiago's own words:

"I have physically traveled in time," says Andrew Basiago, an attorney in Vancouver, Wash. "We have - we did over 40 years ago."

Now Basiago is on a mission - to reveal what he calls a 40-year government cover-up - of Project Pegasus - where he says he was teleported back and sideways in time, dozens of times.

"I have the whole story; I have hundreds of facts," he says. "I can tell you what personnel was at what locations where and which travel device was being used."

And his time travel wasn't recent - it's when he was a kid.

"I entered the program officially in the fall of 1969 as a 3rd grader, age 7," says Basiago.

He says he was one of 140 kids, 60 adults - chrononauts, including his dad, who he says joined him on his first jump.

"My dad held my hand, we jumped through the field of energy, and we seem to be moving very rapidly but there was also a paradox and we seemed to be going nowhere at all," he says.

The TV show "Fringe" aired a similar scene two years ago. A coincidence?

Paradoxes, unscientific claims, unbelievable stories and encounters on Earth and Mars - including meeting Barack Obama when the president was a kid.

Basiago also says he time-traveled six times to the Ford Theatre on the day President Lincoln was shot - but he didn't see it happen. He also saw President Lincoln on another famous occasion, he says. "In fact, during one probe, the one to Gettysburg, the Gettysburg Address, I was dressed as Union bugle boy," he says.

That's right - he claims to have been at the Gettysburg Address. He says a famous photo taken that day proves it. The picture shows a bugle boy who he says is him. It's the only visual evidence he provides for any of his travels - nothing else.

"I was physically at Gettysburg," says Basiago.

He says his time travel experiences show that teleportation – as portrayed on the "Star Trek" series is all wrong.

"No, in fact if you had just arrived via quantum teleportation, the Star Trek method of teleportation, you would have collapsed as a dead person," he says. "And when that bubble closed, we were repositioned elsewhere in time-space on the face of the Earth."

Some would say Basiago is still living in a bubble, but he *has* put his professional reputation on the line claiming time travel isn't science fiction - because he did it.

Of course, it's hard to confirm any of Basiago's claims, but there are over 6,000 websites *worldwide* devoted to him.

Andrew D. Basiago

Something walking on Mars?!

▼

Rat-like creature on Mars?!

The Titor Files

John Titor

2036 Time Traveler

Reminder: some of the following "actual files" (which were collected just for you from sites all over the WWW and left as original as possible) may download oddly, depending upon which eBook reader you use – however, they translated beautifully to print, so you may want to pick up and keep a hardcopy of this book (while sharing the eBook) if you are a lover and collector of weird historical facts that have a **tendency** to disappear over time.

Now let's take a look at one of the strangest and most fascinating tales that's ever been documented – of a modern day Time Traveler.

A man calling himself, John Titor began posting this incredible story on message boards around 1998. *This is NOT a joke.* He claimed to be a Time Traveler from the year 2036.

I've included as many of his conversations as I could find, of which the uncorrupted versions are becoming harder to locate. I will say this about John Titor: if he was a fake – why did he draw so much emotionally charged controversy?

It has been my experience that where there is smoke there is fire – and if this guy and his claims aren't on fire (some twelve years later) – *they are certainly smoldering!*

Some of his forum conversations seem to be repetitive – only because the forum leaders or administrators tried to debunk him by tripping him up.

Try as they did – they couldn't break him! And whether or not you believe – the Titor family *did* hire an attorney who took the job of representing them, and his name is:

Larry Haber

▼

Law Offices of

Lawrence H. Haber, P.A.

Phone: 407 ████████████

Fax: 888 ████████████

(as of 2009)

◊◊◊

At the end of John's forum posts (which have been collated for ease of reading) I've included all the diagrams, pictures of his vehicle, the time machine device, etc. that I could find, and whether or not you believe −

you have to admit, this is damn interesting stuff.

The truth is out there – *believe or don't* – the decision is yours!

Note: the following discussions (which span several years) are long and involved; so grab your favorite beverage, popcorn, or whatever makes you happy – put it on the reader, sit back . . . *and enjoy!*

Conversations of John Titor

November 04, 2000

What does traveling in time look like?

The unit has a ramp up time after the destination coordinates are fed into the computers. An audible alarm and a small light start a short countdown at which point you should be secured in a seat.

The gravity field generated by the unit overtakes you very quickly. You feel a tug toward the unit similar to rising quickly in an elevator and it continues to rise based on the power setting the unit is working under. At 100% power, the constant pull of gravity can be as high as 2 Gs or more depending on how close you are to the unit. There are no serious side effects but I try to avoid eating before a flight.

No bright flash of light is seen. Outside, the vehicle appears to accelerate as the light is bent around it. We have to wear sunglasses or close our eyes as this happens due to a short burst of ultraviolet radiation.

Personally I think it looks like your driving under a rainbow. After that, it appears to fade to black and remains totally black until the unit is turned off. We are advised to keep the windows closed as a great deal of heat builds up outside the car. The gravity field also traps a small air pocket around the car that acts as your only O2 supply unless you bring compressed air with you.

This pocket will only last for a short period and a carbon sensor tells us when it's too dangerous. The C204 unit is accurate from 50 to 60 years a jump and travels at about 10 years an hour at 100% power.

You do hear a slight hum as the unit

operates and when the power changes or the unit turns off. There is a great deal of electrical crackling noise from static electricity.

What are your memories of 2036?

I remember 2036 very clearly. It is difficult to describe 2036 in detail without spending a great deal of time explaining why things are so different.

In 2036, I live in central Florida with my family and I'm currently stationed at an Army base in Tampa. A world war in 2015 killed nearly three billion people. The people that survived grew closer together. Life is centered on the family and then the community. I cannot imagine living even a few hundred miles away from my parents.

There is no large industrial complex creating masses of useless food and recreational items. Food and livestock is grown and sold locally. People spend much more time reading and talking together face to face. Religion is taken seriously and everyone can multiple and divide in the heads.

Will you show us more of the

operations manual for your time machine?

I will consider it but I do not expect they are worth anything to most people except as a curiosity.

If you met yourself on another world line, what would happen?

It has always surprised me why that concept is so hard for people to imagine and accept. Nothing would happen. The universe would not end and there are no paradox problems that threaten existence.

Temporal space-time is made up of every possible quantum state. The Everett Wheeler model is correct. I have met and/or seen myself twice on different worldliness. The first was a training mission and the second is now.

I was born in 1998 so the other "me" is 2 on this world line. There is a saying where I come from, "Every possible thing that can happen or will happen has already happened somewhere".

How is our world line different from this yours?

For starters, the fact that I'm here makes it different. I've also noticed little things like news events that happen at different times, football games won by other teams, things like that.

I would guess the temporal divergence between this world line and my original is about 1 or 2 percent. Of course, the longer I am here, the larger that divergence becomes from my point of view.

What type of vehicle is in the picture you posted?

It's a 1967 Chevrolet.

Does your machine allow you to control time travel?

Yes, it can be controlled. However, the distortion unit has operational limits. Imagine your path through time is through a cone. The farther away from the center of the cone, the more differences you will see in the world line.

The C204 begins to "break away" at about 60 years. This means the level of confidence drops rapidly after 60 years of travel and the world line divergence increases. In other words, if I wanted to go back 2000 years and meet Christ, there is a better than average chance I would end up on a world line where he was never born.

The computer units and gravity sensors "record" your trip and you are quite easily able to return to your point of origin. I am aware that research is being done on faster units with more accurate clocks. I imagine that they will be able to go back farther with a higher degree of divergence confidence.

<u>Are you drawn back to your world line of origin for any reason? Is instability a problem for you?</u>

I'm not sure what you mean by stable. If you mean mentally, there are many things that bother me here but being with my parents right now is important to me. Physically, the only thing really wrong is the number of colds I get.

<u>Are you able to return to your home?</u>

Yes.

Is radiation a problem?

I'm not sure what radiation you mean. If you mean from the unit, it vents X-rays and Gamma radiation out of the rear. As long as you stay away from that, you should be okay. I keep a radiation detector with me to check my environment and make sure the unit isn't "leaking".

What does the light look like when you time travel?

The light bending only lasts a second. Its like driving under a tunnel and being in total black.

Do people know where you are? Can you communicate with your home?

No. They do not know where I am and I cannot communicate with them. Interesting idea though. From their point of view, I will return almost exactly at the same moment I left. From their viewpoint, I will only have aged more than expected.

Is there anything to add about what time travel looks like?

While the machine is on, everything is

black. When the machine is turned off, it is the reverse affect. It appears you are driving out from a bridge. To tell you the truth, I'm usually sleeping when the unit turns off but yes, it does appear that the world fades in from black.

What's the largest technical problem when traveling in time?

The hard part of traveling through time is not the bending of gravity but the plotting of your course and holding to the basic "position" in your environment. This is done through a system called VGL (variable gravity lock).

Basically, the unit takes a reading of the local gravity and samples it during the "trip" in pulses. If the gravity is too far off, the unit stops or reverses itself to the last sample period where the readings were correct. If there is some sort of failure, the unit shuts down and drops out to where ever you may be.

Are there any physical effects from the machine?

The only real physical trace is a large chunk of ground missing from the point of origin and a large pile of dirt at the destination. The gravity field surrounds a small portion of the earth under you and takes it along for the ride. There is really no way around this.

November 06, 2000 09:04

Perhaps it's a bit easier to understand why time travelers do not revel themselves.

Yes, you can travel forward in time. No, you do not need an "invite" from the future. I first saw the car in 2036. The idea was to find a vehicle that would not draw too much attention for the time period. Unfortunately, there were not very many suitable vehicles around in 2036 and I sold the car when I arrived in 2000.

You do not rewrite history. I can only affect what happens here just as easily as you can. Why do people in this time period worry so much about time traveler's destroying their world line when they have no problem doing it themselves every day?

My goal is not to be believed. Most people do not take news of the war very well but I find that everyone believes it's inevitable. Even in your own history, are not great inventions and discoveries made during a time of war in your effort to kill and maim in new and more efficient ways?

No, I do not work for GE or any other company. Are "stock tips" really the first thing you want to know about in the future? As a representative of your time period, do you realize what that says about you? You should probably know that this time is not remembered for its selflessness, charity or ability to work together.

Why would I want to talk to you? Why don't you believe you have something interesting or worthwhile to say to someone in the future?

November 06, 2000 09:08

No, the ice caps are not melting any faster than they are now. There is also far less smog and industrial waste in 2036.

November 06, 2000 16:43

You asked about the North Polar ice pack. I never said the environment wasn't a problem. Doesn't water expand when it freezes? If the polar ice cap melted, wouldn't sea level go down? I don't know if there's enough ice for this to make a difference and I'm not an expert on global warming.

November 06, 2000, 17:26

Yes, I realize people become hostile. I don't expect anyone to believe me and I have nothing to sell. I take no offense by it. Just out of curiosity, if you were a time traveler, what do you think it would take to get people to believe you?

I suppose we could agree that no particular era in history is famous for its development of humanity but just once I would like to hear questions like, "What is family life like in the future? How does society deal with poverty? Is AIDS, abortion and drug use still a problem?"

Why don't I give you a stock tip? The money you make would dilute the intelligence and forethought that a smart person had in picking the stock all by themselves. If I told you how to get rich, I would be taking money from them and giving it to you.

Getting back to my "proper" universe is

tricky but possible. Yes, another jump would take me to a different family.

When traveling to other world-lines there is a system of clocks and gravity sensors in the machine that sample the environment before dropping out. It's called VGL, (variable gravity lock). If a cement block were there, the machine would "backtrack" until it sensed relative congruity to the original gravity sample. A great deal of time and effort goes into picking just the right spot since you cannot physically move during a displacement.

November 06, 2000 22:13

Please keep in mind a couple of points as I answer your questions. First, I am not a physicist. "Time travel" is only a tool that allowed me to do my job 1975. Most airline pilots are probably not aerospace engineers.

Second, let me give you an example of the position we are in. Imagine you live in the year 1900 and a "time traveler" attempts to explain how a jet engine works. Even though the invention of the airplane is only a decade in the future, he would have to find some frame of reference to explain the basics of flight. Then, he would have to outline the mechanics of how the engine works. As amazing as it would sound, the jet would be invented about thirty years later.

Time travel is achieved by altering gravity. This concept is already proven by atomic clock experiments. The closer an observer is to a gravity source (high mass), the slower time passes for them. Traveling at high speeds mimics this effect which = the twin paradox of faster than light travel. However, this type of gravity manipulation is not sufficient to alter your world line.

The basic math to alter world lines exists right now. Tipler first described a working "time machine" through his theory of massive rotating spheres. I apologize for the web site but it was the only one I could find quickly.

Certain types of black holes also exhibit the "time travel" abilities of Tipler cylinders. Kerr was one of the first to describe the dual event horizons of a rotating black hole. As with Tipler's cylinders, it was possible to travel on a "time-like" trip through a Kerr black hole and end up in a different world line without being squished by the gravity of the singularity.

The mass and gravitational field of a micro singularity can then be manipulated by "injecting" electrons onto its surface. By rotating two electric micro singularities at high speed, it is possible to create and modify a local gravity sinusoid that replicates the affects of a Kerr black hole.
For those asking how come a micro singularity doesn't swallow the Earth or want to know details about the size, stability, mass, temperature and resulting Hawking radiation from such a thing, those details I must keep to myself.

Yes, you can travel into the future and it takes less energy than going into the past.

The computer system is connected to the unit through an electrical bus. There are actually three computers linked together that take the same signals from the gravity sensors and clocks. They use a Borda error correcting protocol that checks the integrity of the data and trips the VGL system.

I am not aware of any physical change to my DNA or "aura". I do however seem to be more susceptible to colds.

Yes, you will remember me if you want to. World lines do not change that way and I will only become an insignificant part of your history.

November 07, 2000 17:18

I would equate the "future" GE distortion units to their current jet engines. The first one worked great but they can always make it better. The C204 unit uses 4 cesium clocks. The C206 uses 6 cesium clocks but they use an optical system to check the oscillation frequency. This makes the world line divergence confidence much higher.

November 07, 2000 21:23

Yes I believe in Jesus Christ and we pray to God in churches. There are some differences you may be interested in. Religion is a major part of people's life in 2036. Pain and change tend to bring people together and closer to God. However, religion is far more personal than it is now. There are no huge, centralized religions and people talk openly about their beliefs. It might also interest you to know that the day of worship is Saturday, the day God meant to be the Sabbath and the 10 commandments have been restored to the "10" that God gave us.

Life is much more rural in the future but "high" technology is used to communicate and travel. People raise a great deal of their own food and do more "farm" work. Yes, compared to now, we do work long hours. After the war, my father made a living selling oranges up and down the West coast of Florida. My closest friend raises horses and another works for a company that maintains "wireless" Internet nodes.

Life has changed so much over my lifetime that it's hard to pin down a "normal" day. When I was 13, I was a soldier. As a teenager, I helped my dad haul cargo. I went to college when I was 31 and I was recruited to "time travel" shortly after that. Again, I suppose an average day in 2036 is like an average day on the farm.

Yes, there is a post office. The Internet is still alive and well in the future. People spend more time talking because life is more centered on the community. I've noticed the same type of effect here when the power goes off. People tend to come out of their homes and actually spend time with their neighbors. There is a lot more personal trust and less paranoia.

When I'm with my parents, I live in a community made up of "tree houses" on a large river in Florida. The river floods sometimes and we have access to the Gulf. Most of our neighbors make a living off the sea or in moving cargo by boat.

There is a civil war in the United States that starts in 2005. That conflict flares up and down for 10 years. In 2015, Russia launches a nuclear strike against the major cities in the United States (which is the "other side" of the civil war from my perspective), China and Europe. The United States counter attacks. The US cities are destroyed along with the AFE (American Federal Empire)...thus we (in the country) won. The European Union and China were also destroyed. Russia is now our largest trading partner and the Capitol of the US was moved to Omaha Nebraska.

No new information there (on UFOs and aliens). I find that an interesting subject myself. Personally, I think "UFOs" might be time travelers with very sophisticated distortion units. But that might be a bit wacky.

One of the biggest reasons why food production is localized is because the environment is affected with disease and radiation. We are making huge strides in getting it cleaned up. Water is produced on a community level and we do eat meat that we raise ourselves.

Yes, genetic engineering is used but it's like any other technology. It can be good and bad. One thing we did not do was create more hybrid seeds. What are people thinking?

I am a Christian Agnostic (**HOST**: *Did John mean Gnostic?*). I do not believe faith alone is enough to get us back to God. I'm not sure what happens when we die but I'm pretty sure it's not a walk in the park.

We are working on it (**HOST**: *space travel*).

What future technologies can we look forward to?

I hesitate to answer but I'll give a bit. Hydrogen fuel cells and more efficient solar cells are big deals. Computer technology and software get MUCH better. My greatest joy is sailing. For fun, I enjoy swimming, playing cards, reading, playing games on the net and talking with people who live in other countries. As a community, we celebrate much more and have bon fires and dances. My hobby is sorting through old magazines and videos of life before the war.

After the war, early new communities gathered around the current Universities. That's where the libraries were. I went to school at Fort UF, which is now called the University of Florida. Not too much is different except the military is large part of people's life and we spend a great deal of time in the fields and farms at the "University" or Fort.

Most of my memories growing up are not fond. Life was very hard. Simple things make me happy like hugging my mother and father. Yes, we have cameras. More digital. Film is used like painting is today. No hologram camera though.

Yes we have phones but the service is through the web. Most power generation is localized. It amazes me how much power is wasted now. Yes, solar is big. There is thought that a singularity generator could also be used but most people are against it. The elderly are highly revered and looked after on a community level. So are orphans. There is always something people can do now matter what. The idea of avoiding work is looked down on. Everyone pulls together to keep the COMMUNITY strong.

Hats are more common in the future and flashy colors are less common. Dress is much more functional and we "dress up" whenever we get a chance. I have noticed that no one in this time dresses for occasions even when they have the clothes. Why do people wear shorts to church?

We do not spend nearly the amount of time on our hair as people do now. Women like to wear their hair longer and men have it much shorter. Both sexes shave it all off when they're in active military service.

Far less medical treatment in the future even though It's more advanced. People die when they now its time to die. No lasers. Genetic medicine and cloning organs are the obvious new techs in the future.

The Constitution was changed after the war. We have 5 presidents that are voted in and out on different term periods. The vice president is the president of the senate and they are voted separately.

We have cars, just not a whole bunch of them. There is public transportation from city to city.

November 07, 2000 22:18

If I could bring some material thing back to your time from 2036, it would be a copy of the new US Constitution.

Yes, you could travel to a future that was 0.5 seconds ahead of now but not with my machine. The C204 uses the second as the basic unit of measure. The C206 may be capable of .05 sec.

If you arrived a fraction of a second in my past or future on another world line would I ever know that you arrived?

No "you" would not. But the "you" on that world line would.

It is believed there is some sort of measurable quantum differences in world lines. I am not an expert on that so I can offer little information.

You would be welcome in my home.

November 08, 2000 22:27

Can you explain what the purpose of time travel is on your world line?

In 2036, a great deal of effort is going into "repairing" our environment. I was sent to 1975 to get a computer system and take it back to 2036. Time travel is not a secret in 2036 and I expect it will become more common.

Why is time travel used?

Right now, its used to get information or "items" that would be helpful in getting a post WWIII world back to a normal condition. There are 7 other time travelers in my unit.

Where do time travelers the most?

Right now, most of our practical missions are from 1960 to 1980. There is a great deal of research into later and future periods but the farther you go, the lower the divergence confidence of the world line.

You said there is more than one time machine. Are they all being used?

Yes.

What type of vehicle will you get to go back in since you sold the other car?

It's a 1987 4WD. The vehicle needs a strong suspension system to handle the weight of the distortion unit.

Are you able to take other people with you when traveling in time?

Yes.

November 11, 2000 18:46

If you change cars do you have to re-adjust the time machine?

Yes. But it's a function of the VGL system. A gravity baseline is taken and rechecked every time the unit is used. A new vehicle would alter the gravity signature.

What type of system is used to maintain the singularity?

I am not a physicist so I cannot answer that to your level of sophistication. The singularities are held in an enclosed magnetic field.

Can you dimensionally travel with the time machine?

No. However, the longer the unit is on past a safe divergence confidence, the "stranger" the world line becomes. The unit I have is safe to about 1% for every sixty years at max power.

November 11, 2000 18:56

If your time machine is accurate to the second then you must have a reason for choosing the moment you arrive.

My goal was to reach a certain date and time which is converted into seconds for the computer to make its calculations. I do not pick the second. It's more important to have a low divergence confidence number.

Are you here now to observe the results of the presidential election?

I would use the word "elections" a bit cautiously. Perhaps it's easier now to see a civil war in your future?

November 11, 2000 19:07

If you're here to get a computer, why are you are posting on a message board? I think you're committing treason!

Why would it be treason?

November 11, 2000 19:11

You're time machine travels at 10 years an hour? That would take you more than three hours? How can the air trapped in your vehicle last that long?

Yes, that's about right but my initial trip was to 1975, not 2000. I guess it's a question of how many technical details you really want or you feel I'm making up. We do take additional 02 and the air pocket is a bit larger than you might think.

November 12, 2000 16:41

Do you worry about anything happening to the people that help you here? What if someone tries to build a time machine based on what you've said?

Yes, I have considered it but it is very easy to remain hidden behind a veil of disbelief. The people who understand what they are seeing are not aggressive. Everyone else just finds them entertaining. The obvious first answer is that the only world line of consequence is my own and I don't care what happens here. That however, is not the case. I have shown these documents in order that people might consider the possibility. I do not expect people to believe them.

What is the real reason you're posting these top-secret government documents?

The restricted nature of the documents I posted refers to the departmentalized nature of the technical information. The manual is supposed to remain with the unit. The current F-16 fighter jet probably has an operations manual. All the information in it is not secret but no one wants it to "walk" away from the plane.

Did your commander authorize you to post this information or is it something you did on your own?

I am here for personal reasons. For a few months now, I have bee trying to alert anyone that would listen to the possibility of a civil war in the United States in 2005. Does that seem more likely now? Actually it's quite amazing to see what's happening. I have been trying to get people to pay attention for the last few months but to see it unfold is very interesting. Before I leave, I'll try and post my report.

I am curious... will anyone be upset if Florida's votes are not counted in the Electoral College because of the current "confusion"?

Have you thought about the possible outcomes of your actions?

Yes.

12 November 2000 16:41

Have you considered the dangers for someone who helps you here while you remain covert? What about the people that read the information and might attempt to build a time machine?

Yes, I have considered it but it is very easy to remain hidden behind a veil of disbelief. The people who understand what they are seeing are not aggressive. Everyone else just finds them entertaining. The obvious first answer is that the only world line of consequence is my own and I don't care what happens here. That however, is not the case. I have shown these documents in order that people might consider the possibility. I do not expect people to believe them.

November 15, 2000 13:56

A broken clock tells the right time twice a day.

November 15, 2000 14:12

Your vehicle looks new.

Thanks. I sold it when I arrived here. It attracted a great deal of attention. Perhaps that's why it was still around in 2036.

You must be in the army.

More or less correct.

Your commanders must have larger time machines they can use to come and find you.

Why would they come looking for me? I'm expected back but I will only have been gone a few seconds from their point of view.

You should probably know that our government will find you.

Stupidity and greed are fairly predictable for a period of time.

If you're telling the truth, the last thing you should be doing is talking about the war and the government.

Have you considered that your society might be better off if half of you were dead?

Why don't you tell us everything about the stock market and future before you go?

I'm thinking about it. I'm waiting for my family to buy up all the good stuff cheap first...(joke).

November 15, 2000 14:20

I want to believe you. In order for me to do that, I need a prediction that comes true right now.

I appreciate the position you are in but you must realize that I am not affected in the least if you believe me or not.

Tell me who wins the (upcoming sporting event).

Off the top of your head, can you tell me if it rained in Atlanta this time last year? Do you think time travelers carry a sports reference with them?

Tell me anything like that and I will believe you forever.

It is a mistake to give anyone your

unwavering belief...but you will find that out yourself in 2005.

November 15, 2000 14:33

Can anyone operate your time machine? Does it have a key? You would be welcome at my home.

The unit has two security systems to protect it from "most" people. One, it has a code that must be entered correctly. Second, and probably more effective now, the unit cannot be used by anyone who can't add and subtract.

November 15, 2000 14:41

Why did you go to 1975?

The first "leg" of my trip was from 2036 to 1975. After two VGL checks, the divergence was estimated at about 2.5% (from my 2036). I was "sent" to get an IBM computer system called the 5100. It was one the first portable computers made and it has the ability to read the older IBM programming languages in addition to APL and Basic. We need they system to "debug" various legacy computer programs in 2036. UNIX has a problem in

2038.

November 17, 2000 09:34

It is thought that being close to a gravitational field has a biological effect on all matter including cells. The effect is to slow the movement of electrons in the orbits of their nucleus, which slows the mechanical and biological functions of the observer close to the gravity. Thus the passing of time is a local phenomenon depending on how close you are to a gravitational source.

This is one example of a theory involving "time shells" progressing in size and intensity around a gravitational point from all matter. The more massive the object, the larger and more influential the time shells around it (like an onion). Another offshoot of this theory is that kinetic energy is actually the conversion of stored energy in the atom as it passes through time shells in a gravitational field.

November 20, 2000 17:16

You sold your car? How will you get back?

The unit is portable between vehicles. It is very heavy and requires a "stiff" suspension. The unit is currently in a 4WD.

How did you solve the normal problems of living in this time?

I am currently living with my parents on this worldline. They know exactly who I am and how I got here.

How did you get access to a scanner and computer?

There are numerous ways to do that. Any local printing shop allows you access to a computer system.

With all the things you could be doing here, why are you trying to find out about us on the Internet?

What suggestions do you have?

It seems like it would be hard to be from the future and fit in on this timeline.

You don't think you could blend into 1970? What difficulties would you expect to have that couldn't be overcome?

There are a lot of people here who think you are crazy.

I have nothing to sell nor do I want anyone to believe in me or take some action. What other people think of me does not affect me in the least.

If telling us about your time machine won't change anything, what would happen if someone built one based on your information?

What you do on your world line is your own business. I can't think of any better way to start a war than for someone to figure out how to make a time machine. Go for it.

I'm not trying to ruin your reputation or credibility; I just want straight answers to my questions.

I will be happy to answer "almost" any question you have.

November 21, 2000 10:41

All of the questions asked have been

answered in one way or another. You assume I am here to start a war?

Consider this: You are a time traveler who wishes to go back in time to 1941 because your grandparents live close to Pearl Harbor in Hawaii. You realize you can't stop the war but you may be able to help them prepare for it. Strangely, December 7th comes and goes with no sneak attack. As the war in Europe rages on, Japan fails to join the axis power, there is no war in the Pacific and the United States remains neutral. Then, you watch as Germany begins to develop the atomic bomb... all by themselves.

For a change, I have a question for all of you. I want you to think very hard. What major disaster was expected and prepared for in the last year and a half that never happened?

As far as war goes, I have faith you are quite capable of starting one all by yourself. I am hard pressed to accept any criticism on my outlook on that subject. Growing up might have been a vastly different experience for me than it was for most of you. Personal responsibility, determination, honor, friendship and self-reliance are not just words we try to live up to or fantasize about.

On my world line, life is not easy. We live in a world recovering from years of war, poison, destruction, and hate. All of it, courtesy of the thinking and actions of people that live right now in the same world you do, worrying about which stocks to buy or whether or not a stranger is lying to them on the Internet.

I believe that hardship and challenge develop character and community. My first experience with war came when I joined a shotgun infantry unit at the age of thirteen. In the 4 years I served as a "rebel", I watched hundreds of people get shot, burn and bleed to death. I know exactly where I was and every detail of the exact moment the first nuclear warheads began falling on Jacksonville. I know the pain and regret of not acting soon enough to enjoy a relationship as a loved one dies of brain cancer from a war that gained nothing.

How can you possibly criticize me for any conflict that comes to you? I watch every day what you are doing as a society. While you sit by and watch your Constitution being torn away from you, you willfully eat poisoned food, buy manufactured products no one needs and turn an uncaring eye away from millions of people suffering and dying all around you. Is this the "Universal Law" you subscribe to?

Perhaps I should let you all in on a little secret. No one likes you in the future. This time period is looked at as being full of lazy, self-centered, civically ignorant sheep. Perhaps you should be less concerned about me and more concerned about that.

November 21, 2000 21:31

There have been a great many questions piling up that I do plan to get to. If anyone sees a question I have answered before, please feel free to repeat the answer. I do not plan on leaving this world line for a while yet. I very much enjoy spending time with my family.

November 22, 2000 20:58

The observation of time travelers "appearing" suddenly in a world line does not happen very often. There are two cases and two points of view to consider.

In the first case, the time machine does not move as it goes from one world line to another and then returns. The people watching on the original world line would wave good-bye and watch as the machine is turned on. There would be a static discharge and the air would appear to "ripple" as if it were getting denser. Then, it would stop and the machine will have appeared to have disappeared. If the machine doesn't move its position from world line to world line, the observer would not see it disappear at all.

In the second case, if the machine were moved, it would disappear from the viewpoint of the observer and return in a different location based on where it was moved and turned on from the destination world line. In that case, the rippling seems to dissolve the machine and it disappears. If that happens while you are watching it leave and you expect it to return, you know it was moved or had a serious malfunction.

It is actually quite dangerous to get too close to a distortion unit as it enters or leaves a world line. It vents radiation and has a very strong localized gravity field. Personally, I worry about that a great deal.

It can be adjusted to some degree (the gravitational fields). The CG (center of gravity) is adjustable within about 4 feet and the unit is effective about 10 to 12 feet in either direction from there. The vertical distance is quite a bit shorter and is determined by sensors in the unit.

Depending on whether or not you are going forward or backward, the footprint of the unit is different. I wouldn't quite say it "scoops" up the ground cleanly. It sort of vibrates it loose and takes it along for the ride. It looks like someone raked the ground an inch or so deep with a small hand hoe or shovel. The negative ergo sphere "scoops" up the front and back areas of the field. The positive ergo sphere leaves a longer area near the center of mass. It's about a cubic foot of dirt spread out over six square feet or so.

If your device was inside a building and used, what would happen?

It might not be as destructive as you think. Depending on how close any object is to the field, it might not do any damage at all except for the floor.

What would happen if something touched the distortion field as it is turned on?

It would be quickly spread out over the lateral length of the gravity field. Imagine being squished and stretched at the same time. I would imagine anything left after that would be vaporized and generate static electricity.

What is the temperature around the time machine during operation?

Very! Hot. Depending on the power setting, 100 to 120 degrees is average.

What position is the car in when the time machine is being used?

The car is off and the brake on.

Has the time machine ever been used while it was moving?

Not that I'm aware of. Its important that it remain as still as possible so the gravity sensors can get a good lock. The divergence confidence would be way off if the vehicle were moving.

Do you wear a uniform?

I wear a standard civilian service uniform during instruction and training. It's sort of a cross between an army uniform and overalls. We do have a quartermaster who distributes clothing appropriate to where ever we are going.

There is a patch. It is round and has a graphic of a Kerr singularity (sort of looks like an eye with gravity waves around it) with two spiral paths running through its center. One path represents the "safe" way and the other is the path to God. The bottom of the patch has my unit number along with "Temporal Recon" printed on it. However, we remove any identification and patches before we go anywhere.

How long will you be on our world line?

I'll be around for a while yet.

November 25, 2000 09:10

I must apologize for the poor quality of the information. There's a running inside joke about the technical issues. If the unit has a serious problem its not as if anyone can use those drawings to take the electron manifold off the singularity housing with a flat head screwdriver.

What is number seven in your drawing?

That sensor detects various parameters from the singularity.

Why can't we see the atomic clocks?

There is another page that depicts the computer and clock systems. That technology is not new and not very interesting.

There is a cut-a-way drawing of the entire unit that I will probably post before I leave.

November 25, 2000 13:57

UFOs are as much as a mystery to me as they are to you. Would you be surprised to know that there are a great many people who don't believe in time travel on my worldline?

I do believe UFOs are a real mystery but I also know that chaos theory dictates that no matter what technology or knowledge we have, there are always unknowns.

November 25, 2000 13:59

On my world line, it is known that the 5100 series is capable of reading all the IBM code written before the widespread use of APL and Basic. Unfortunately, there are none left that anyone can find on my world line.

November 25, 2000 14:01

It is quite difficult to get used to some of the communication patterns I have come across here. Confusing conversational conflict for anger seems to be a common and typical problem. Why does the expression of differing emotion seem to threaten so many people? I do not feel accused nor was I trying to accuse anyone. Your opinions are as valid as anyone's and I do not suggest you change them because of anything I say. I never said I was here to start a war although I have strong opinions about what a war would do for you. I am not aggravated by words.

Imagine you are Jewish and you are able to travel back in time to Germany in 1935.

All around you are the patterns of thinking and action that will lead to a great deal of harm, death, and destruction in just a few years. You have the advantage of knowing what will come but no one will listen to you. In fact, they think you're insane and the situations you describe could never happen.

What I feel is not anger, it is sadness that you cannot see what I see.

November 25, 2000 14:03

I appreciate your offer for help but I am quite all right. The responsibility for the "disaster" is your own but I do not consider it a disaster. Rebirth is often painful. My world line is not unified under a single government but I would say it is closer to a unified purpose. Isn't that what you want anyway?

No, I do not have the ability to make calculations that would affect world lines to my advantage. Besides the fact that manipulating people for personal gain is wrong, I am of the belief that it is best to make the world line you are on now better.

I like the incredible freedom you have on

this world line but I see it as a trap and I fear the cost is the loss of your sense of connection with family and community. Yes, you can self actualize your ambitions but at what cost to the people around you, or yet to be born? The incredible availability of art, literature, and limitless resources is hardly taken advantage of as I imagine the number of people sitting in front of their TVs.

No, I have not tried any fast food. Thinking about where the food came from, how it was shipped and treated absolutely terrifies me. I have tried to tell people about CJD disease and it seems to be "catching on" in Europe.

Technologies themselves are not lost but some of the older tools and techniques have been lost. I think there is more detailed information about the war posted earlier. The energy stored in the singularity is used to create the distortion fields. That energy is created in a particle accelerator. I'm not sure what you mean by "temporal turbulence". What effect would that have that would need to be overcome?

When I leave, I will return to 2036. The

computer I have is expected there. I am unaware of any "true" inter-dimensional device available on this world line now. I would image the effects of such a device would be hard to hide.

My "ship" is not broken. I am here by choice. Don't you find current events interesting? Again, I'm not sure I understand your last question. Perhaps we should agree to the definitions of a few terms and basic physics before I try to answer that.

Can you tell us the foods that are unsafe now? Is there anything we can do to prepare for the war you are describing?

I tried to consolidate your questions into a basic list. I hope this helps.

1. Do not eat or use products from any animal that is fed and eats parts of its own dead.
2. Do not kiss or have intimate relations with anyone you do not know.
3. Learn basic sanitation and water purification.
4. Be comfortable around

firearms. Learn to shoot and clean a gun.

5. Get a good first aid kit and learn to use it.
6. Find 5 people within 100 miles that you trust with your life and stay in contact with them.
7. Get a copy of the US Constitution and read it.
8. Eat less.
9. Get a bicycle and two sets of spare tires. Ride it 10 miles a week.
10. Consider what you would bring with you if you had to leave your home in 10 min. and never return.

What event started the war? Can it be stopped?

The war is a result of faulty politics and desperation from Western leadership during the US civil war. Yes, I suppose you could stop it.

Are some areas of the United States safer than others?

Take a close look at the county-by-county voting map from the last elections.

Were biological or chemical weapons used in the war? Were any weapons used that effected people's minds?

Yes there were biological and chemical weapons used. No mind control weapons but there are new "non-lethal" weapon systems that turn out to be quite lethal.

Has cancer or AIDS been cured yet?

Aids, no. Cancer, some progress.

What would you say to any world leaders who might be reading this right now?

Revel in your confidence today because you will not win tomorrow.

What is the one thing you would want us to remember?

Please, please wake up. Look at the signposts around you now.

Earlier you spoke about the patch you wear on your uniform. You said the spirals represented two paths.

One was safe and the other lead to God. What does that mean?

The safe way is the one calculated to take you in time where you want to go, the other path will take you to God (death). Both are equally accepted and anticipated before each trip.

December 2000

December 06, 2000 21:36

What is music like in 2036?

Most of the same music you enjoy is available on archive. However the days of mega-stars playing multiple track studio produced music and lip-synching on a huge stage are pretty much isolated to your time period. Like everything else, music is much less centralized. The general trend is away from "computer generated" music and more toward real people playing real instruments. Music is much more of a personal experience. More people know how to read music and play together in small groups.

What are the social prejudices in 2036?

Yes there are. However, as odd as it may sound, it serves a useful purpose in my time. First, you must realize that your experiences with "prejudice" and mine are different. I would characterize the intolerance you have here as a result of ignorance and fear. I have observed that people with unfounded and irrational fears about their fellow man in this time have the luxury of not having their beliefs tested.

After the war, much of the prejudice you have now was swept away by simple necessity. People had to work and fight together just to survive. This has a way of opening a person's eyes as to the value of fellow human beings.

What difference does the color of a man's skin make when you are both fighting against the same enemy to survive or find water or grow food? On my world line, if a man doesn't pull his weight in the community, then we feel prejudice towards him as a burden to us. This feeling of shame he experiences then makes him realize his responsibilities.

What is the health care like in 2036?

You would probably not like it at all. I would

compare it to what you see in Western movies. We do have hospitals but there are more family doctors and house calls as compared to what you are used to. There is no real organized health care. If you get a serious disease, you die. This however has had a tendency to strengthen the general genetic pool. Doctors are more concerned about things like broken legs, snakebites, and delivering healthy babies.

What is the entertainment industry like in 2036?

Again, entertainment is less centralized. There are "movies" and "TV" but everything is distributed over the net and more people produce their own "shows".

What's it like to start from nothing and work your way up in 2036?

Very easy. If you can work with your hands and get along with other people, you will always find work.

Could you travel in time to escape your original worldline? Would you be looked down on for that?

From a physical standpoint, I suppose that

would work if you were trying to escape a natural disaster. But you must consider that trying to "escape" by time traveling has it's own problems. I'm not exactly sure what you mean by being looked down on.

Is there some sort of new world government in place by 2011?

On my world line in 2011, the United States is in the middle of a civil war that has dramatic effects on most of the other Western governments.

Earlier you said something was wrong with the UNIX computer code on your world line. What's wrong with it?

Yes...and with yours too. I have to believe there must be a UNIX expert out there some place that can confirm that. I'm not exactly sure what the technical issue is but I believe some sort of UNIX system registry stops in 2038.

Does the continuing conflict between Arabs and Jews have anything to do with the upcoming war?

Yes.

When you first started posting online, did anyone believe you?

I don't know if anyone believes me now. But you must realize that I don't expect anyone to believe me or does it affect me one way or the other. I enjoy these conversations too. What would you expect a time traveler to say or do in your time?

How does the singularity stay cool in your time machine?

Very highly technical, cooling stuff. [sorry, I don't get much of a chance to be a wise guy]

How far have computers and software progressed in 2036?

Good question! I would say the biggest difference is in the reliability of the hardware and software. It absolutely amazes me how willing people are here to accept computer and software failures on such a regular basis. I was very surprised to see how prolific it is. You can look forward to very stringent manufacturing parameters and programming discipline. Think back to the early days of the computer and how much work and cleverness it took to fit those programs into such small areas of memory. Has more and cheaper memory brought better programs or just more programs?

As far as technical specifics, I'm afraid I cannot go into too much detail. However, I will tell you that processor speed and memory size take dramatic leaps forward.

The question involved very complicated physics with magnetic fields, faster than light travel and tachyons.

Hmmm . . . I afraid my notebook on Maxwell's equations isn't with me and I must say again that I am not a physicist. As far as tachyons goes, it's my understanding that they cannot exist at all unless they are created and "traveling" faster than light. As far as time travel goes, I'm afraid tachyons are merely useful for describing various quirky effects of strange matter.

I will post again as soon as I get a chance.

December 10, 2000 11:00

I will try to answer all of your questions however; I have noticed that there seems to be a difficulty in entering a conversation in the middle. Most of the following questions I have answered at length in previous posts. I understand how important it is to have these answers when new people read portions of the posting but I find it tiring to keep repeating myself. Any suggestions would be appreciated.

Who won the Super Bowl in 2001?

I do not answer questions like this. Although I don't really know the motivation for the question...I can guess. If a time traveler had knowledge of your future, and you could only ask one question, would this be it? Besides, can you tell me if it rained in New York on June 4th 1932? You are from their future so should you know that?

The question involved how the time machine could travel to Earth in the future or past since the position of the planet would change in space. The question also asked about life support during travel and how physical structures are avoided.

This is actually a very good question that parts and pieces of the answer are scattered around in previous postings. I am often surprised that it is not the first one asked.

You are correct; this problem is actually the most difficult part of time travel. Although some of your assumptions about matter displacement are a bit off, the problem is real. Inside the displacement unit are a series of very sensitive clocks and gravity sensors. This system is called the VGL (variable gravity lock). In simple terms, before the unit "leaves" a world line, it takes a base reading of the local gravity and adjusts the Tipler sinusoid to "lock" into that position. Although the temporal physics of this statement are wrong, in effect, it holds you to the "Earth". During travel, it periodically checks to see that the field has not varied. If it does, it stops and reverses course or drops out at that point. Buildings and other terrain features are avoided in the same way. Yes, we do bring oxygen in the vehicle with us but we do not lose atmospheric pressure.

If the time machine was used for a split second and then returned, wouldn't it appear someplace in far off space as the Earth continued in space?

Please see VGL system above. Also, please be aware that from the viewpoint of the time traveler, the displacement unit actually left and then returned. The viewer does not experience this. Think of it as a Gosub statement in a computer program.

December 10, 2000 11:32

What type of money do you use in 2036?

Its not very different than it is now. Yes, we have money and credit cards. However, like everything else, the monetary system is decentralized. Banking is based mostly around the community structure. There are no multinational banking or computerized economic systems. There are also no income taxes.

Are radio location systems still used on livestock?

Some of the larger farm communities use electronic markers and fences. I am not a farmer so I don't have very many details about that.

Is the GPS system still in use in 2036?

Oh yes! In fact, the unit I have with me works here. I'm not sure why that surprises me. There are also a great many communications and Internet satellite systems.

Do you still have an American flag? What does it look like in 2036?

Yes, we still have the flag. There is a debate about changing it from 50 stars to 5.

Do any of the states have new names?

Not that I can think of... No major cities anyway. I am aware that some smaller towns changed their names after the war and most universities have the word "fort" before them on my world line.

How much training is required before you can time travel? Can you make any repairs to your time machine?

I started training on the recommendation of my PS officer when I got out of school in 34. I graduated in March of 35. Actual hands on training started soon after that and I left 2036 in April. No, there are not very many repairs I can make to the unit. The unit is built very well but the tolerances are very small. I could probably repair the electrical systems and it will accept inputs from older computers.

Is it still safe to fish in 2036?

Yes, we can fish. There are some areas that are still too dangerous to spend a lot of time in so we can't fish in those areas.

Have any time travelers entered your world line in 2036?

No I am not aware. But I accept the possibility.

Is there public transportation in 2036?

A high-speed train system connects the larger cities. Roads are still used for cars and many people ride horses and bikes.

What items will you take back with you to 2036?

Books! I'm also taking copies of family photos that were lost during the war.

How long have you been on our world line?

That I must keep to myself right now.

What music and books do you enjoy?

Do you play an instrument?

I read a great deal of history. I like piano classics and some older rock and roll. (why did that type of music go so far down hill?) No, I don't play any instruments myself.

Don't you worry about our government finding you?

Not really. In order to find me, they would have to believe time travel is possible.

Are you taking pictures and video?

Oh yes.

What is your favorite food?

It's very hard for me to find food here. It all scares the Hell out of me. I've found a couple of local farms where I am reasonably sure the raw food is safe and my mother started a garden for me.

Do medical advances in 2036 have anything to do with genetics?

Again, I'm no expert. I believe there is a great deal of progress in treating the cancer cells with modified viruses. So I guess the answer is yes.

The question involved details on natural disasters and temperatures in 2036.

That's one area I've decided not to talk about, sorry. The average temperature worldwide is a bit cooler.

Are any other time travelers from 2036 with you now?

No. They are off doing something else I'm sure.

Is time travel discussed on the web in 2036? Do you recognize any web sites now from the future?

Time travel is a major point of social discussion. I'll have to check when I get back. Most of the larger servers that were in the major cities were lost or destroyed. I'll try to bring a copy back with me. Perhaps you can check that in your 2036.

Is there any connection between you and the John Titor that was here on this worldline before you? What do your parents tell him about you?

He calls me uncle. Yes, there is a connection. He feels like a younger brother. Sometimes I get mixed emotions watching him and realize I'm watching the origins of my personality. I tend to playfully criticize my father about that.

December 13, 2000 12:44

Well...it looks like the election is almost over. I have been quite busy in the last few days and I appreciate everyone's patience. I should be thanking you for listening and even if everything I have said before means nothing, I expect it has at least been entertaining. I went through the postings looking for questions I have not addressed yet. I hope I got to all of them. I get my email remotely through about three different systems and I use one of your hand-held computer units to write with. In a few days, I would like to send Pamela a list of questions of my own. As you know, one of my areas of expertise is in history and the information I have gathered has been invaluable. Although I am not leaving right away, I would like to include some of your thoughts in my report.

<u>Isn't it possible someone who reads</u>

this may invent time traveler? When John leaves, no one will remember him and this will all be forgotten.

This statement is quite insightful. One of the reasons I have been communicating this way is so that others that do not post who I have directed here can see some of the information I have shared with you.

Won't even a small divergence between world lines cause everything to look different to you as compared to your worldline?

The divergence measurement refers to the local gravitational field as compared to the point of origin. It is merely an empirical indicator of overall change in a world line. Some things that are quite different on one world line have very little effect as time passes and the worldliness appear to converge" again and look very similar. World line changes are not exponential; they act more like chaotic attractors with varying effect depending on their size and location.

If multiple world theory is correct and there are two of you here, one of you has to be the real one. Also, I find

it unsettling that you look at us as representative of the people from your past. One reason I think you may be a time traveler is because I don't understand or relate to your thinking at all.

You are correct in your thinking. This world line is not mine but it looks very similar. It may be compared to reading two books that are the same. I can jump back and forth between them and still see the same story. I do not consider you true representatives of my world line but I do know something of how the story ends. In consideration of Everret-Wheeler, the reason we don't know if Schrödinger's cat is dead is because it might have had a time machine.

I think Russia is still very likely to attack the United States with nuclear weapons. It's hard to imagine being here to see it.

You are also correct but I want to add a twist to your thinking. Russia's enemy in the United States is not you, the average person. Russia's enemy is the United States government.

If this world line is 2 percent

divergent from your world line, how do you get home? If you go forward from here to 2036, won't the divergence approach infinity?

Yes, this is true. If I go forward on this world line, the future will not be my future. I get home by going back to 1975 before I arrived and then going forward to 2036.

Your deductions are quite accurate. It's possible to go forward to "your" 2036 and it would look nothing like mine.

I have a working IBM 5160 computer. Maybe I should stash it away for thirty years and see what happens.

Toss it. The 5100 is the interesting machine.

Does the current relationship between Arabs and Jews have anything to do with the coming war?

Real disruptions in world events begin with the destabilization of the West as a result of degrading US foreign policy and consistency. This becomes apparent around 2004 as civil unrest develops near the next presidential election. The Jewish population in Israel is not prepared for a true offensive war. They are prepared for the ultimate defense. Wavering western support for Israel is what gives Israel's neighbors the confidence to attack. The last resort for a defensive Israel and its offensive Arab neighbors is to use weapons of mass destruction. In the grand scheme of things, the war in the Middle East is a part of what's to come, not the cause.

What can you tell us about (Obscure Physics Question)?

Perhaps you could describe what you mean? It appears to be space travel or the effect of similar atomic particles that seem to be related over great distances. Again, please forgive my ignorance; I am not a trained physicist.

Does anything happen in the year 2012? I've heard stories about the world ending.

In my 2012, I was 14 years old spending most of my time living, running, and hiding in the woods and rivers of central Florida. The civil war was in its 7th year and the world war was three years away. Yes, there are unusual events in 2012 but they do not cause the world to end. Unfortunately, I have decided not to discuss events that you or I can do anything about. It is important that they be a surprise. Perhaps you are familiar with the story of the Red Sea and the Egyptians?

The question lists a number of paranormal conspiracy theories and then asks about weather control in the year 2036.

I must admit I am unfamiliar with most of what you have asked here. I am aware of the Harp system but I don't know how they would control weather with it.

Please describe any details you can about the education system in 2036.

The education system has been through many changes. School in 2036 is no longer a political indoctrination system and students "learn how to learn". Since community activity varies from place to place, the emphasis on basic reading, math, and language is augmented with skills particular to the community. One school may emphasis farming while another teaches woodworking. Having children is a bit different and less common in 2036. A typical school day involves a setting very much like it was 100 years ago with smaller classes and few administrators to teachers. Other areas of study that are less common now are history, citizenship, and personal economics.

Are you going to take your family away with you to protect them from coming events? Are you worried about somebody finding your time machine?

No, I am not taking them with me but I am trying to prepare them for the future as a promise to my Grandfather in 1975. I am not really that concerned about the "time machine". It is quite safe.

How and why do the Arabs Jews become entangled in the civil war of the U.S.A?

They are not directly involved but political situations are dependent on Western stability, which collapses in 2005.

The Arab countries appear to have weapons of mass destruction. Do they use them against America?

Not against America but they are used against each other.

If you could post a picture of money used in the future, that would be more impressive than your time machine pictures.

I am disappointed that you feel I am trying to impress you. Why would I bring money from 2036 with me? Besides, isn't that something that could easily be faked? Now if I told you I was your cousin's brother and I knew about that scare on your left leg...that would convince you.

Before you leave, you could videotape your departure and get it to us?

That is an interesting idea. I will look into how to do that.
    1. According to the Constitution, who do you think has the final word on choosing a President and why?
    2. Do you think the Electoral College should be continued?
    3. Why do you think the Bill of Rights was written?
    4. What is your opinion of the second amendment to the Bill of Rights?
    5. Does anyone know what industry is the largest US political contributor?

    6. Does anyone know who owns the

majority of the solar power research and development companies?

7. Imagine you have all the money and power you desire. What do you see yourself doing?

8. What is the longest distance you ever ran and/or walked at one time without stopping?

9. What is the longest period of time you went without food?

10. Do you know what Cholera is and how to avoid getting it?

December 21, 2000 10:59

Very interesting argument but I have a couple of questions. You described the word paradox as, "...it refers to the "idea" of the existence of a problem that has no solution." Actually, the #1 definition I read in the American Heritage Dictionary is "...a seemingly contradictory statement that may nonetheless be true".

Also, what exactly is your definition of "time travel"? I was taught that time travel is strictly a local observation that can only be measured by the experience of an individual or single particle. Under that definition, the "twin paradox" (time dilation due to acceleration or gravity) and even sleeping can be considered time travel. You appear to be arguing against dematerialization and/or space-like trips under the limits of special relativity in a single world line.

I do agree that the "grandfather paradox" is not possible simply because the classic problem is presented as an observer's issue magnified to a universal issue. Your statements about observation are correct when you isolate the experiences to a single world line.

However, the reason there are no paradoxes is because the universe doesn't care how we react to its handy-work. In a Universe made up of infinite worldliness (super universe), everything is possible and has a 100% probability, therefore...no paradoxes.

Where do you think all the water came from in the great flood stories?

I believe the explanation for the "great flood" stories originate with the changes that occurred near the Mediterranean at the end of the last ice age. Even on this world line, there is a great deal of evidence to support the fact that sea levels did change radically in isolated areas worldwide.

I also heard some place that if the ice mass on Antarctica melted today, sea level worldwide would rise about 100 feet. I'm not exactly sure that's true but still... Mt. Everest might be a bit of a stretch.

I do however agree with you that there are no physical paradoxes but for the opposite reasoning.

Peace to you also.

30 December 2000 11:47

Greetings and happy holidays everyone. I am very surprised and delighted to see the conversation going in the direction it has on this thread. Unknowingly, you all have stepped into the real mystery of time travel that remains speculative in 2036.

Based on a couple of questions I see here, I will try my hardest to describe what we in 2036 think space-time looks like and how it behaves. Please keep in mind that I realize how easy it is to dismiss what I say. First, I'm trying to do this from memory. Imagine you are back in 1911 trying to explain a jet engine to the Wright brothers.

However, there are some very basic

properties of quantum theory that support this model today. I appreciate the fact that you are reading this with an open mind.

If parallel universes do exist, did they all start simultaneously?

It is thought that the event called the "Big Bang" was the start of not only this world line or universe but all worldliness and all universes that make up the super universe. It is also thought that the super universe can be imagined as an expanding sphere with the big bang in the center. Individual worldliness (or timelines as you call them) can be imagined as lines originating at the center and "trending" toward spiraling around the sphere until they reach the edge.

The individual world lines expand in length and widen as you follow them from the center. Each individual "moment" or "event" on a world line has infinite possibilities or outcomes. Imagine this as a single point with infinite lines shooting away from it, which in turn, are made up of points with their own possibilities and outcomes.
Now, remember, these individual worldliness with all these points and possibilities are defined by their ability to

hold there inhabitants to time-like trips only (no faster than light travel).

Now consider the reality of a spinning or electrified black hole (Kerr). Penrose diagrams of these oddities show mathematically that you can make simulated space-like trips (faster than light) through the singularity without being destroyed.

In order to do this without wiping out most modern physical laws, you must travel to an alternate world line or universe. Therefore, if multiple world lines exist, infinite world lines exist.

In trying to imagine a super universe with infinite possibilities and world lines, I think of a room with mirrors on all the walls. You are aware of your captivity but as you look in the distance, you see an infinite number of "yours" in an infinite number of mirrored rooms.

The gravity distortion machine allows you to "step" out of your room and into another next to you. The closer you are to your original room, the closer it looks like yours, the farther away, the stranger it looks to you.

If I go forward on this world line, the future will not be my future. I get home by going back to 1975 before I arrived and then going forward to 2036.

A few people have asked me about this statement so I will try to clarify it.

On my world line: (A) in 2036, I was given a mission in 1975. I turn my machine on and jump to another world line (B) in 1975 with about a 2% divergence from (A).

From the very point I turn my machine off on (B), I create a new world line just because I am there. This line can be described as (C) and started when I got to (B).

I am now doing my mission on line (C) in 1975 when I discover a very a good reason to go forward on (C) and see what happened. I turn my machine on and go forward on (C) to the year 2000.

When I turn it off, I start another line called (D). So from my perspective, here we are on line (D) in the year 2000. In order to go home to line (A) I must turn my machine on and go back on (D) until I reach (C) which in turn would take me back to (B) which in turn takes me to a point before I arrived on (B) then I go forward from the point I arrived on (B) back to (A).

If all this isn't enough to get your head spinning...here are some issues we're dealing with in 2036.

1. Did your world line (D) exist at all before I got here from (C)? (personally I don't see how it couldn't).

2. What happens at the end of a world line at the edge of the super universe?

3. If there are infinite world lines and infinite possibilities and an edge to the super universe, doesn't that mean occurring events on worldliness are staggered as they reach the edge? (time could end at any moment without warning).

Happy New Year everyone!

December 30, 2000 10:28

<u>I think you're taunting us with our ignorance on the questions you've asked on the Constitution.</u>

It would be nice to be able to remind everyone about their rights and responsibilities but I am not here to judge you. I am not capable of that nor would I want that in return. As you know, my interest is in history and in the paradox of thought. I do however, find it interesting how important the Constitution became to the average US citizen's life, if even for a short moment.

Young people should look forward to the future while older people don't see many days ahead of them. I think it's pointless for you to ask us to worry about politics if we are all doomed.

It saddens me that you do not realize your true worth as a keeper of information and experience. Perhaps the end that we fear will open your eyes to your true value as an individual. Young people need wisdom. The captain of the ship knows where the lifeboats are.

There are numerous experiments going on at CERN. In order for them to make a black hole they would have to travel faster than light.

I'm pretty sure they have a number of experiments going on at the same time at CERN. The one I'm referring to involves very high energies using protons. From my historical perspective on my world line, I do recall the issue was a point of contention about 18 months ago or so.

There were some scientists who thought the experiments were too dangerous to try. The time travel I refer to does not require faster than light travel and due to multiple world "reality", paradoxes do not occur. Natural time machines do exist.

30 December 2000 13:17

I fear our conversation is in danger of turning (to conflict) due to an effect that is quite common on these boards. I realize what I'm saying is quite hard to swallow and it causes debate, weather serious or entertaining. It is even more difficult when you come into the middle of a conversation or a series of questions that are a few weeks old.

Your points are all quite valid and I have discussed them at length on this and other boards for quite a while. I do not wish to antagonize you however, we both know the

Tippler cylinder is only a thought experiment to explain the very real physics behind Kerr black holes. As to your other comments, again, they are all true as defined by the limits of space-like trips on single world lines. It does not account for travel between world lines.

I have never claimed to be a physicist or an expert on what the CERN laboratory is doing at any given moment so I feel it is pointless to argue about what they may be doing in the future or what "breakthroughs" they will or might have. My comments about the CERN lab are in reference to particle accelerators in general and other questions that have come up in the past. The major physics break through for controlled gravity distortion does happen at CERN in your future. Heck, we haven't even touched on "Z" field compression yet.

I suppose I could say that I was the one that traveled in time and convinced them to change their experiments but even I would have a hard time believing that one and I do not wish to insult your intelligence.
30 December 2000 13:37

Thank you for trying to answer those questions (from 30 December 2000 11:47) but I really do not expect that anyone can. I thought I would share with you things we wonder about. Your logic about me is quite correct but again I must state that I am not trying to get you or anyone else to believe or buy anything.

As far as evidence goes, I have however decided to try an experiment with you that may be more convincing. It involves the travel of information at faster than light. In fact, I have dropped at least three little gems like this that no one else has picked up on.

You said you are confused by the 5100 story. I will explain further. In 2036, it was discovered (or at least known after testing) that the 5100 computer was capable of reading and changing all of the legacy code written by IBM before the release of that system and still be able to create new code in APL and basic.

That is the reason we need it in 2036. However, IBM never published that information because it would have probably destroyed a large part of their business infrastructure in the early 70s. In fact, I would bet the engineers were probably told to keep their mouth's shut.

Therefore, if I were not here now telling you this, that information would not be discovered for another 36 years. Yet, I would bet there is someone out there who can do the research and discover I am telling the truth. There must be an old IBM engineer out there some place that worked on the 5100. They just might not have ever asked if I hadn't pointed it out.

30 December 2000 23:26

I apologize for wasting this much space but I thought some of you would be interested in seeing this after reading some of things I've been saying in the last few months. Below is the address to the news site and a copy of the text.

31 December 2000 12:43

I'm flattered and a bit overwhelmed. I can honestly say I've never quite had this experience before. I appreciate the news posting. Thank you.

31 December 2000 11:00
Well, you're getting closer people. Here's another one I found today. Again, I apologize for taking up this much space but I thought you'd want to see this.

January 2001

03 January 2001 13:47

I've been reading the last few postings with a bit of confusion. I see there is controversy over my "story" that is causing some people to ask themselves if they believe it or not. For quite a while, I have been stating that not only do I not expect anyone to believe me, it's irrelevant and in my opinion, quite dangerous. Belief implies that you accept what I say as true and real. Over the Internet, this is impossible. In fact, I have stated before, there are many people in 2036 who do not believe in time travel.

I also think that unwavering belief is dangerous. One very disturbing thing I have noticed about your society in general is your blind acceptance of what you are told. Do you really think the news industry doesn't have an agenda? Do you really think those hamburgers you stuff into your body are safe? Do you really think your government is telling you the truth? What proof do you have of any of that?

What I do want you to do is open your eyes to the events that happening around you that have nothing to do with me. Some of you have been reading for a while now about the war in 2015 and the breakthroughs in particle physics that would be coming soon. Doesn't the CIA report on 2015 and news on the z-field compression at least support what I've been saying a little bit?

I just saw another story today about the Russians moving Nukes into the Balkans to thwart any future expansion by NATO. I also haven't heard anyone take me up on my "information experiment" on the IBM 5100 or check out the information I've given you about the UNIX failure in 2038. With all due respect, I find it hard to take some of you seriously.

05 January 2001 13:46

In 2036, community life is a bit different. People are valued and judged based on their contribution and worth. Work is organized around the family and the value of that work is assessed inside of the community. Most communities range in size from 1000 to 4000 people. If a family wanted to move from one community to another or if a son or daughter wanted to move to another community, they must apply and be interviewed by the community leadership council. During this process, the family or individual is evaluated as to whether or not the work or skill they have is required or necessary to that individual community. Once accepted, the family or individual is expected to uphold their end of the work and support the community. If they don't, the community stops supporting them and they are forced to change their attitude or move away from the community.

The family work we did was picking, sorting and shipping oranges by sailboat up and down the coast of Florida. We were expected to produce a certain amount for the community and a certain amount for other communities as agreed to by our CLC. In exchange, we received power, water, a certain amount of food and other necessities that were produced inside our community.

I see this message board as a small community and I have no other way to value the contributions of others on it other than what my past experiences tell me. I have tried to answer as many questions as I can without being annoying, repetitive or inappropriate, and for some of you, entertaining. Under these conditions, I have decided to seek guidance from all of you, the other members of this community, as to whether or not my postings are of any value to the direction of these discussions. If they are getting distracting or repetitive, I will stop and continue to enjoy reading your thoughts and ideas.

Who receives the Nobel Prize for time travel? Since you claim no two world lines are the same, this information could not hurt anything.

There are a great many people involved with the discovery of time travel. Just as I will not give "stock tips", I will not divulge their names as that may impact their lives now.

What would happen if you gave us technical information that's discovered in a year or so?

If I had any and I published them, I'm sure they may have a large impact. Unfortunately, I don't have any with me. Even if I did, I'm sure they would be scrutinized. Again we get back to the same question. If you were a time traveler, what would you do to establish your credibility?

Have any other time travelers visited your future in 2036?

No, I am not aware of time travelers visiting my world line in 2036. However, that does not mean it can't or isn't happening. Also, the possible number of world lines a time traveler might arrive at would place the chances of them hitting any particular one at very long odds.

What happens to (name of person asking question)? Am I involved in Time Travel? What of the (armed) resistance?

I have no idea what happens to you in your future. There was a resistance on my world line but their goal was to maintain power and control over other people. We killed most of them by 2020.

How big is the IBM 5100?

I would say its about 20" long, 10" high, and 30" long.

Timetravel-0 and I are not getting along.

I'm not aware we had a falling out. I apologize if you think that's the case.

Are there still books in 2036?

Yes, books are still published.

Are there restrictions to what you can do with the time machine?

The displacement machine is not mine but I am free to make certain decisions based on the experiences and information I gather from each world line. I am expected back but from their perspective, I will only have been gone for a split second.

06 January 2001 13:10

What is the progress on string theory?

Who doesn't love string theory? Please forgive the next few comments, I'm trying to be cryptic and jump starting my memory at the same time. In 2036, string theory still dominates physics due to its continued "effect" of encompassing other physical properties from unrelated fields.

A great deal of the theoretical mathematics behind time travel was discovered by testing various ideas in string theory and eliminating the anomalies. As I recall, it was this original work that led to the final proof that six dimensions do indeed curl up to give us our observable universe. This in turn supported more of the theoretical math behind time travel.

It's ironic that the beauty of string theory gives future engineers the confidence to create the distortion unit even though the final proof is still unknown. You're a physics student, have you ever heard the Princeton String Quartet play?

09 January 2001 09:28

Why don't you give us more detailed information about the future so we won't be so skeptical?

I think skepticism is a good thing and no one should lose it.

Why don't you accept my challenge? What harm could it do?

I'm not sure what "challenge" you are referring to. If you mean the live chat, I have no problem with that. I do that quite often on other boards. However, I fear I have very few bread and circuses left and I fear I am becoming quite boring. Also, I'm not sure I fully understand the nature of the challenge.

You could still give us details about the future that wouldn't harm anyone and still add to your credibility.

Again, I do not seek to add to my credibility. There is no point to it. Actually, by providing information that was useful, I would be adding to your collective fear that I am real. In that case, this cycle we are in concerning "truth" only spirals and gets worse.

Why can't you name the five presidents you have in your time?

Over the past few postings, I have tried to describe the limits of what I will talk about and why. Here is a short recap list. In future postings, I will place the following number next to each question as to why I will not discuss it.

1. I will not disclose any information that will cause someone to personally gain by its knowledge. This means no stock or sports tips.

2. I will not disclose any detailed information that would allow someone to avoid death by probability. This means no earthquake or bombing information.

3. I will not disclose any information that may compromise any future actions by individual people or threaten their family and well-being.

On my world line, we are no longer afraid of the "NOW". Are you afraid of Nazis?

The reason the job of President was split into an office of 5 has 4 main reasons. With 5 (presidents), foreign policy is more consistent, power shifting between parties has less of an impact on the overall government, individual strengths between presidents add to the strength of the overall office, and one president is elected for each major area in the United States.

What is the extent of Presidential power in 2036?

The office of President is far more diluted and decentralized than it is here. The powers of the national government are more defined and reside more at the county and state level.

What is the new government like in 2036 compared to the current one?

I think the new government is good.

However, since the concept of nationally subsidized welfare is gone, most people here may not appreciate it.

Where is the new US capitol?

The new US capitol is in Omaha Nebraska.

How do the 5 presidents take power?

The voting for individual candidates is on a rotating schedule.

What are the other time traveler's destinations, and missions?

I am not aware of the details of other missions. Of the seven, three had already left before I did. I suspect they are on similar missions.

Are any other time travelers on our current world line?
No, the chances of that are very slim.

Are you in contact with other time travelers?

No, although I would suspect that is not impossible I have no idea how you would do that.

Is it possible to send a message through time?

Unless the information physically travels with the person; none that I'm aware of.

09 January 2001 12:24

I apologize for being unable to respond faster. I looked back and couldn't find any questions but I may be mistaken.

I also see that you seem to have a great deal of anger directed at me. If you knew where I was and how to get the distortion unit, what would you do? What judgment and punishment should be passed upon me?

On this and many other discussion boards are people who are genuinely interested in and hoping to experience or discover time travel. What punishment would you apply to them if the succeed? I see that you too have had time travel experience. What punishment do you deserve?

When time travel was discovered, there were many people who were against its development. However, once the true nature of time was realized, the resistance faded. Even if one world line was able to ban, kill, and stop all time travel, it will continue on another.

However, the corollary is also true. Take heart, on some world line, you have succeeded and all time travelers are dead by the hand of your followers and thinking or talking about it is a crime.

Perhaps we should all agree what we mean when we use the phrase "time travel". We are all moving forward in time just by existing and the effects of acceleration and gravity do "slow" time down for the observer so the rest of the world seems to speed up around them.

If we call this natural time travel (time-like trips on a single world line) then perhaps its really controlled time travel that you are after (space-like trips on a single world line or traveling to alternate world lines).

January 10, 2001

With that, it may then be useful to separate the concept of controlled time travel into theory and practice. In theory, time travel was taken seriously by mainstream science when it discovered that Einstein's equations do indeed support the possibility of controlled time travel under special relativity. Since special relativity (under its current limitations) has been proved useful and accurate in predicting other physical phenomena, it is widely believed that controlled time travel is also possible.

Nearly all of the solutions that allow time travel in special relativity also have the annoying problem of crushing the time traveler in a wake of radiation and gravity.

In my experience, there is only one safe way to obtain controlled time travel and that involves the "safe" properties of a Kerr singularity (black hole). However, I do not discount the possibility of other methods either physical or metaphysical. I'm just not sure I would bet my life on them without any math to back them up.

10 January 2001 01:13

$E(8) \times E(8)$?

10 January 2001 23:10

You mentioned a divergence percentage between time lines. How is it possible to measure divergence?

The measurement for world line divergence

is an observation variable isolated to the distortion unit. An effective analogy would be a "gravity radar". The unit's sensors take a "snapshot" of the local gravity around the unit before a flight. During travel, this baseline is periodically checked to make sure there are no major changes in the environment that would cause a catastrophic mass failure (brick wall appearing from nowhere). The percentage of VGL divergence from one world line to another is a calculated guess by the three computers that control the unit based on its starting point. It is useless in describing characteristics of individual world lines.

There is a bit of folklore about the first distortion driver who reaches a destination with a zero divergence. This would mean they had traveled on a space-like trip to their own world line of origin. This paradox is quite possible although highly unlikely. I wonder if anyone out there can take current string theory and make that one work on paper?

You said 6 curled up dimensions. The current string theory suggests that there should be at least 7.

I may be mistaken but I thought it was

pretty well established now that (N-10) was on track.

11 January 2001 09:04

I must admit (forum name), you have succeeded in confusing me. The more I read your postings, the more I question my understanding and local use of social interaction, courtesy and logic. However, I also believe that all viewpoints have some inherent value even if it's not apparent.

I take it you won't answer my last post?

I took a look at your last posting and didn't see any questions. However, I now realize that I may be mistaken in assuming much of what you wrote was rhetorical. In may be helpful if you add some indication that you are asking a question you wish someone to comment on.

If you're from the future, than you should know what happens to (personal names of people on forum)? Am I involved in time travel?

This is the only question I found from your postings that you could be referring to.

Again, I have no knowledge of you in any possible future nor would I comment on it if I did. As far as the photograph of me in a uniform, I'm not sure what that would prove even if I had one.

If you speak to me anymore, I will probably discredit your claims of being a time traveler.

I do not seek credit for anything. The most I hope for (for the most part) is to be at least interesting and engaging.

If you want meet my challenge (to talk online), I await your answer.

Perhaps I was unclear before. I did provide the web link earlier that does have a chat room set up for time travel. My schedule is a bit more flexible right now at least for the next week or so. Please take a look and let me know when you will be there (open to anyone of course). Since there really is nothing to be gained or lost, I look at this not as a challenge but more of an opportunity to get to know everyone better.

You need to answer people more quickly. You answer questions only when it is convenient for you. Your

answers appear to be just enough to satisfy someone but you have enough time to just look it up in a book.

I'm not sure what you mean by that. In earlier postings, I have stated that I'm trying to avoid repeating myself and frankly there are some items that are just over my head or that I have no knowledge of. It is curious that you feel knowledge can be something owned or somehow that becomes less worthwhile if it's passed on.

11 January 2001 11:49

I don't believe I ever said I came back looking for a UNIX bug fix. I came back for a computer system. Don't you find UNIX useful now?

Temperature is about the same although there were anomalies after the war.

Just curious.... what does everyone think of "IT"? (Ginger)

15 January 2001 12:04

In the post that follows, I've tried to answer the latest questions directed at me but I am hoping you all may be able to add some insight into something I've noticed. In our attempt to communicate here, some of the comments on this board have become increasingly hostile and negative. I see the same type of interaction when I watch news interview programs.

The guise of productive interaction and communication is thwarted by illogical verbal attacks and misdirection. I understand why the news does it. They are trying to hold an audience by generating conflict. For a while, I thought that was the goal here too but it appears that anger and conflict is being created on this site to cause genuine harm and pain. Its hard for me to believe that this is being done on purpose so I have concluded I simply do not understand some hidden element of your collective social interaction.

On the other hand, if its being done for no reason, I would understand a little better how people in this time could accomplish so much and yet be so vulnerable to their emotions and fears. I think it was Thomas Jefferson who believed that the only way to sway opinion was through calm, respectful, intelligent conversation.

Weather I'm a time traveler or not, I

suppose there are numerous ways to view my "story". By the nature of the communication medium, I believe it's impossible to prove therefore it's impossible to believe. I agree that conversation spurs ideas. If I'm not a time a time traveler, than perhaps the seemingly disjointed statements I make will actually create the idea in one of you that leads to "real" time travel.

What is a WORLD LINE?

Individual world lines represent the limits and paths physical objects take through space-time under the laws of special relativity. They can be shown graphically on an x-y graph with x representing distance in space and y representing passing of time. In time travel talk, world lines are used as a way to describe and separate the experiences of a time traveler because various laws of special relativity appear to breakdown and can't be defined on a single world line. (The word) World line has also become synonymous with "alternate universe" and / or "time-line".

You've admitted to taking things back with you. Do you pay for those items?

I pay for these items with money. Personally, I believe stealing is wrong.

Couldn't you just take what ever you wanted back with you?

There are mass limits to what can be taken back.

Are you tested mentally before becoming a time traveler? What makes a good time traveler?

Yes, there are numerous psychological tests. I was chosen based on my educational background and military service. The training lasted about two years. There is a great deal of physical training to counter the physical effects of distortion. They were also looking for drivers who had a fair amount of self-sufficiency and an ability to function under extreme isolation and confinement.

What do they look for when choosing someone for a mission?

Depending on the mission, time travelers are usually chosen for a particular mission based on their ability to gain the cooperation of someone related to the goal on the target world line. In my case, my grandfather was directly involved with the building and programming of the 5100.

If a time traveler is bad and commits murder or goes crazy, is there anyway they can be recalled?

There is a difference of philosophy between us that should be clarified. Since I believe that all possible outcomes and events are possible, probable and certain, it is impossible to assign "goodness" or "badness" to a person or situation.

On some other world line, I am an insane time traveler causing destruction and death while (name of forum member who confronts John on a regular basis) chases me with his band of devoted followers. However, on this one, I am not. Since both events are certain, their value is neutral.

You can only assign goodness and badness to the events and experiences you have direct control over or witness. Only actions are good and bad, not people or things. I suppose if I was a psychotic killer, I could accomplish my mission, avoid capture and still return to my worldline of origin without penalty. There would be no way for them to know what happened. However, I believe that action is wrong and I would be accountable to my God.

What exactly is "IT" since no one will know for sure until probably 2002 according to news reports?

I suppose this question is my own fault. As a time traveler, I am expected to know every winning horse and hot stock as well as the weather in all parts of the world at all times. I was genuinely interested in your opinion of how "IT" was being presented and advertised. Do you feel manipulated? Do you think it's really a big deal? Do you like the way the news is dealing with it?

How do you know our world line and yours will follow the same path?

This world line and my own are almost exactly alike.

15 January 2001 13:36

Thank you for considering the problem of (my) returning home. You seem to have stumbled on an intuitive proof of some of the physics of time travel. You are correct, getting back to the world line of origin is easier than picking an exact destination on a different world line.

I wrote down the graphic you outlined. If y_1 starts perpendicular to x_1 and x_2 and is rotated, where is the center of rotation? I imagined it between x_1 and x_2. If this is so, wouldn't y_1 end up parallel between x_1 and x_2 with each one being 6 inches away from y_1 on either side?

Does the micro-singularity that powers your time machine physically reduce the size of objects during operation?

Actually, there are 2 singularities in the unit. The gravity field is manipulated by three factors that affect it in distinct ways.
Adding electric charge to the singularities increases the diameter of the inner event horizons. Adding mass to the singularities increases the area of gravitational influence around the singularities. Rotating and positioning the polar axis of the singularities affects and alters the gravity sinusoid.

The effects of the gravity produced by the unit do not have enough time to significantly alter physical objects within a reasonable distance from the outside of the sinusoid. No, things do not get smaller.

If the electron injection system alters the shape of the field, would that not force the unit to accelerate through space as well as time?

There is no relative movement in space due to three main factors. Large, kinetic energy inducing effects of the gravity field are compensated for by the interaction of the singularities. The mass of the unit and any objects inside the sinusoid do not exhibit any huge increases on the departure world line during travel. The observed path of the traveler is obtained by changing the gravity, not by moving the vehicle. The black hole comes to you.

The question is define "time".

To me, time has two definitions. I see time as a mathematical component of a 10 dimensional super universe. It is a variable I use to define my location and existence.

I also see time as a metaphysical compromise our senses use to define the area of collective existence God has placed us in.

When I can measure and sense time, I know I am not with God.

16 January 2001 10:43

<u>**I know you have a hidden agenda. You have not made clear your motives and are only giving us very little proof. Why are you holding back?**</u>

Rest assured there is nothing I have planned in my "hidden agenda" that will make anyone's life any worse than it already might be.

<u>**Am I right? You do a lot of work in the background of our society? Admit it.**</u>

I plan to leave soon. There is nothing I can do here that will affect my home. My goals are based on the love I have for my family. Actually, my interest in the past is a result of going through piles of half burned books and magazines left over from a war started by people you share this planet with right now. On that note, perhaps its more interesting to consider what I won't be doing to try and stop that war.

<u>**What are your personal beliefs? Do**</u>

you believe in God, which one? Do you think the ends justify the means? Don't you have people regulating Time Travel?

How many Gods are there? I believe in just one. What are your suggestions for regulating time? Perhaps a list of your "time rights" would be something to spark conversation on the board.

You are a cheater of life. Why can't your world line fend for itself? You take the solution from us and threaten to change billions of lives (here). Why don't you care about that?

I'm just guessing that "cheater of life" is not a compliment. I can't think of a reason why my actions would be immoral and someone else's would not. You suggest I'm capable of changing lives. I suggest that I'm no more capable of that than you are.

I'm just not so sure you recognize your own potential. You don't need a time machine to save or destroy people. If there was another person doing the exact same things I was but they didn't have a time machine, would they be putting billions of lives at stake and suffer your judgment?

Is there any such thing as wrong for you?

Yes, I believe in wrong and right. I judge my actions based on God's law. Is it wrong for me to murder? Yes it is. Is it wrong for me to teach someone how to defend themselves and they commit murder? No, I am not their keeper. Since I am leaving, I will be incapable of causing any harm. But, what damage will you do with anything I have said?

17 January 2001 11:29

In the future, sociologists spend a great of time discussing the collective mentality before the war that led to the demise of "Homo Materia". Many of them point to an experiment that was done in the 1970's or 1980's. The experiment isolated various sizes of rat populations in varying cage sizes with varying food and cleaning schedules. It was discovered that no matter what, there was a certain ratio of rats to space that once overtaken by population would always lead to aggressive and destructive behavior in the rats until enough of them had died or been killed to get back under the ratio. This was true even when the rats that were given plenty of food and had their cage cleaned every day.

Besides the occasional school or office shooting and violent video game, I can't help but think about that experiment every time I see someone stranded on the highway or walking on the side of the road carrying an empty gas can. I ask my parents why we don't stop and help and they tell me they are afraid of being attacked and of the possible consequences of helping someone they don't know. I would respond by pointing out that it's our duty to help someone, not just because it's the right thing to do, but because we can never know that person's true worth and the risk of losing then is too great.

I didn't fully understand their stubbornness until I saw a news story about a doctor who was sued for applying emergency first aid to an accident victim who died. I believe your society is biologically geared for self-destruction. However, I feel strongly that does not excuse me from my responsibilities as a temporary member of this community.

Although (name of poster on forum) is a bit quirky and eccentric, he does belong to us.

He's this community's quirky and eccentric guy and although he can be aggravating, I can't help but feel protective of him. As I'm sure most of us believe, under all of his postings, he has some interesting things to say. When I first read one of his postings, I first shook my head and then I began to question my own understanding of not only the English language but of the real meaning of the odd things he brings up. If it's done on purpose, it's quite effective.

26 January 2001 15:11

After trying to post a couple of times and seeing the number of postings increase but not the pages, I assume it is limited to 11 pages. Any thoughts?
26 January 2001 18:46

Greetings everyone. I've been away for a while so I apologize for not getting back to these questions sooner. Since (personal name from forum) brought up a few things I had addressed in the mysterious mail that never made it to the board, I will post it here. The others are responses from other questions.

Why are you so interested in the Constitution?

After the war, the United States had split into five separate regions based on the various factors and military objectives they each had. There was a great deal of anger directed toward the Federal government and a revival of states rights was becoming paramount. However, in their attempt to create an economic form of government, the political and military leaders at the time decided to hold one last Constitutional Congress in order to present a psychological cohesion from the old system.

During this Congress, the leaders discovered and decided that coming up with a new and better form of government was nearly impossible. The original Constitution itself was not the problem it was the ignorance of the people that lived under it.

Don't you like your "new" Constitution?

From my viewpoint, it's very effective. I am a very strong believer in local or state's rights.

When are you leaving to go home?

There are certain windows I must wait for in order to leave. There will be two this year.

The first one opens this spring.

Is it possible for you to return to this time line once you leave?

Not with the machine I have now.

If we all had time machines, would we end up destroying the rest of the local world lines?

Since everything is already happening and possible on different world lines, the answer is yes, and no.

You appear to be a Libertarian who stresses the need for mankind to get back to basics. You also seem to be a gun rights activist.

I suppose from your vantage point that's a fair assessment. I would call myself more of a centrist. Although I understand the "gun rights" issue here, I cannot relate to it all and it is a common point of argument with my mother.

I keep saying her tune will change in about ten years and she'll be cleaning shotguns in her sleep but it doesn't help. If it makes you feel any better, I never shot anyone who wasn't trying to kill me.

Since we're not as smart as you, can you help us solve the technical problems of time travel?

When you say "us", what do you mean? Do you mean "you"? Where would you go if you had my machine? How do you think the rest of the world would react to the U.S. having a time machine and they didn't?

You stated you went back in time from 2036 to 1975 with a near 2 percent divergence. You also said that a zero divergence is a myth or technologically improbable.

Yes, a "ZD" is thought to be impossible. However, consider that an exact entry point "may" not be necessary to get home. The important factor is the path, not the destination. Under multiple world theory, there are an infinite number of "homes" that I could return to that don't have me there. The divergence for that window is somewhere near .0002377%.

Please correct me if I don't understand. Are parallel universes created by events in our own world

line?

Parallel universes exist independently of each other and only interact to avoid the collapse of the wave function for any given particle or event that you are looking at.

I like to imagine it as a series of parallel lines crossed by a wavy wave. Each point on the wavy line where a straight line crosses it represents an alternate outcome. The multiple "yous" on each world line record a different result for the activity of the particle.

Multiple World theory and parallel universes have never been proven. It only exists as a way to explain various findings in quantum mechanics. Just because people talk about it doesn't mean it's real.

I agree with you that an explanation doesn't make it so. **However,** you can build a model to describe physical behavior. Even if the model is not complete, its "truth" can be measured by how well it predicts the behavior it describes.

Where did multiverse come from as a theory? Where is the evidence of its

existence?

I believe the closest non-related evidence for multiple universes right now comes from the physics (derived from special relativity) of rotating (Kerr) black holes. If you examine a typical Penrose map, science agrees that you can travel to "other universes" through these cosmic oddities. They can't be different places in your own universe (world line) because you would have to violate the speed of light limit to get there.

Since the existence of multiple universes is a reality from my viewpoint, please allow me to disclose an idea we toss around a bit in 2036. Since all possibilities, outcomes, and events are occurring and exist simultaneously, it would mean there are multiple universes out there where "you" are living a day behind and a day ahead of the "you" on this universe.

There are some who believe that memory is some sort of information transfer or communication with the "yous" in the past, across world lines or universes. Although this is seemingly quite ridiculous, if you think that could be true, than physics tells us that the same information transfer from our future selves on other world lines is not only possible but certain. Could it be that fantasy or "what if" scenarios are actually future memory from an alternate "us" on a future world line?

According to physics, there is no reason why this cannot be true.

26 January 2001 18:51

I now see it plainly states the topics are limited to 11 pages. I guess this only proves I'm not the brightest singularity in multidimensional space-time.

26 January 2001 21:21

Your mission to go back in time seems rather pointless since you stated earlier there is no way to get back to your exact starting point.

The reality of infinite possibilities is rather difficult to get a grip on and if it were not for the math, I would delegate it to the realm of religion. I like to think of it as standing in a room with mirrors on all the walls. I can look to my right and left and see many "mes" all doing exactly the same thing. If we all took a step to our right and passed through a dimensional doorway to the next mirrored room, it would be very difficult to tell if anything had changed. In that sense, there are an infinite number of world lines waiting for me to return with the computer. If I can get to one of them, I have completed my mission.

It would seem then that another time traveler that looks like you could arrive on your world line of origin with another IBM computer and no one would know the difference.

Bingo!! Seems like something they would do a lot of psychological testing for before they sent us off.

26 January 2001 21:28

<u>Where does it say it is limited to 11 pages?</u>

I was just trying to be clever. However, I am still unable to see any postings past 412. Can everyone else see them?
John Titor.

01-27-2001 12:45 PM

Greetings. I am a time traveler from the year 2036. I am on my way home after getting an IBM 5100 computer system from the year 1975.

My "time" machine is a stationary mass, temporal displacement unit manufactured by General Electric. The unit is powered by two, top-spin, dual-positive singularities that produce a standard, off-set Tipler sinusoid.

I will be happy to post pictures of the unit.

I have been communicating online with others who are interested in time travel.

28 January 2001 09:45

And you never showed us a picture of your uniform.

Actually, there are numerous places I have posted pictures. I believe the links are still on the board.

Why haven't you commented on not caring about your original world line?

I'm not sure what you're asking. I think those statements speak for themselves and your interpretation of them may be, unique.

28 January 2001 12:22

One of the reasons I like this board so much is that the questions are more thought out, the people seem to be a bit smarter than normal and I'm not continually bombarded with questions about stock tips. I will admit that on a conceptual level, you are picking it up much faster than I did.

If you leave your world line, some of your friends and family members will never see you again. That seems like a very immoral thing to do.

I'm not sure why you think it would be immoral. Don't soldiers today go on duties they may not return from?

It depends on how you define what the real "me" is. If you consider the mirror example again, as all of the "mes" step one room to the right, the family and friends in that room (and the time traveler for that matter) would not be able to tell the difference.

The probability of us noticing a difference is based on the divergence of the trip. If all events and outcomes are certain, there are world lines where I do return for every world line I don't return to. All the "moral" events would then balance out to zero. Again, it's hard to judge good and bad outcomes, only good and bad decisions.

What's the point of fixing something in one world line if there's an infinite number of world lines where a problem doesn't exist?

Our actions and decisions are based on the knowledge we have in our own world line. Yes, the bell shaped curve is a useful tool but if we are capable of change for the better than we feel we should at least try.

Even if "I" don't return to my exact world line, a similar "me" probably will. Besides, I just look at it as helping a world line where their time traveling me didn't show up but I did.

Earlier you said the people on your world line would only experience your being away for a split second. How can this be true if you'll never return to that world line?

Again, I refer to the mirror example.

It seems impossible that another you would return to the world line you

originally came from.

I'm not sure I said another time traveler "will" return, I think I said they "could" return. The location/gravity "map" I have of my path getting here could be duplicated with a fairly high degree of accuracy. It's just that my machine was not designed to do that.

I think it's a mistake to rely on the concept of the impossible when dealing with the reality of multiple worlds. Keep in mind there are an infinite number of "yous" on infinite world lines having completely different experiences with "me".

You stated that returning to your world line of origin is thought to be impossible.

My "exact" ZD world line that is.

01-28-2001 06:35 AM

My initial flight was from 2036 to 1975 (61 yrs.). I then went from 1975 to 2000 (25 yrs.) Later this year, one of two favorable windows will open and I will return to my 2036 (35 yrs.) I am here now for personal reasons. The web page is not mine. I have been speaking online for about three months and the page is a collection of the various documents and pictures I have sent to other individuals.

Also, I realize there is no way for anyone to believe me with absolute certainty so I hope I'm at least entertaining. You may be interested to know that even in 2036, there are a large number of people who don't believe in time travel. Are you sure the world is round?

01-29-2001 07:47 AM

Please keep in mind the web site is not mine and I apologize for the poor quality of the files. The photo you saw was taken by me with a Polaroid camera manufactured here. The other documents were duplicated by placing a book onto a copy machine at a packaging and shipping store and then scanning and saving them.

As for the printing technology in 2036, you may be surprised at how many people use typewriters however I agree the documents were probably not created that way.

I too am very anxious to hear your thoughts and questions on time travel / gravity displacement and any comments on the Everett Wheeler Graham model.

01-29-2001 03:48 PM

Although the documents posted were printed from a computer printer, is it really that hard to believe that manual typing is just a bit more common in thirty years? After the war, many things like manual printing machines, bicycles, sailboats and hand tools were valued a great deal. I have noticed more people in California are installing wood burning stoves.

I realize my claims are a bit ridiculous but my intent is not really to be believed. However, if I had an opportunity to talk to a time traveler, I might ask questions like: How exactly does the singularity sensor measure the expansion of the inner event horizon or why does the reality of multiple worlds support the religious dogma that there are no good or bad people just good and bad decisions or what were the political motivations that changed the U.S. Constitution?

In my experience, when it becomes necessary to convince someone what I do for a living the only way to do that is to be related to them. Everything else is immediately written off as a parlor trick, even if they're standing in front of a cooling distortion unit and I show them a dollar bill with the year 2029 on it.

In the last few months, I have had numerous extended conversations online and there are quite a few things I've said which can easily be checked out but haven't. I get no pleasure out of being right when it comes to CJD disease, war in the Middle East or suffering people in far away lands. There's nothing like the look on someone's face when you tell them 100,000 people will be dead tomorrow. In my travels, I have discovered that most people really don't want to know about the future because if its different than what they want it ticks them off. Actually, I don't blame them.

The means by which I travel in time is very physical. I require a "machine" to do it. It weighs about 500 pounds and gets quite hot. I do not own it and I did not build it. Within limits, I will be happy to discuss how it works and how "future" science thinks time works. No we have not completed string theory yet but (N-10) seems to work pretty well.

As far as the future goes, your world line is about 2.5% different than mine. This is a roughly cumulative measurement based on my arrival in 1975. As far as I can tell right now, you are headed toward the same events I would call "my history" in 2036.

However, the very nature of time travel states that every world line is unique and you are very much in control of what you do and how you get there. Heck, the fact that I'm here makes it different from mine.

I have nothing to sell, and there is nothing I want anyone to do. For all other time travelers out there, I have no tests for you and I would enjoy discussing your feelings and experiences after the war. To everyone else, while I'm here, I am very interested in your philosophy, religious outlooks, and speculations on technology.

29 January 2001 12:25

Are the Olympics still being played in the future?

As a result of the many conflicts, no, there were no official Olympics after 2004. However, it appears they may be revived in 2040.

Why wouldn't it be possible to use that recording to go back to your original starting point without any divergence at all?

Perhaps it's better to say it's so highly improbable as to be considered impossible. A good example is the concept of trying to get closer to something by cutting the distance in half for every step you take. Since the computer is basically making calculations from an imperfect model of reality, there are no absolutes. I also believe there is a theory that states you would have to violate the speed of light limit to have a perfect ZD.

Why are there only two windows of opportunity for you to go back? Does this have anything to do with the weather?

The weather isn't a factor as much as gravitational tidal forces are at the point of arrival.

I disapprove of war because I think it's immoral. What do you think?

I disapprove of murder. Man as a species is

incapable of changing his nature through will alone and war is a tool of biology. The ability for war sleeps in each one of us and we must decide what we will do before the beast awakens. As for morality, again I point to the "universal" balance of good and evil. For every world line where there is peace, there is a world line that has destroyed itself.

What's the use of bringing back the computer to a slightly different world line since you know there are just as many world lines in which no-one returns to bring back computer?

The decisions and actions we take as individuals can only help those who we have direct interaction with. I believe it is wrong to be capable of helping and do nothing. My struggle is in the irony that if everyone just "did nothing", on every world line, there would be no action and thus no immorality or evil (no good either).

I'm very interested in your comments on the greed of humanity. You're unique perspective of humanity at this point (2001 and 2036) is valuable.

Please feel free to ask anything you like.

Can you record a voice message for us before you leave?

Yes. I am considering trying to videotape my departure and having my parents post it after I leave. That should keep you all busy for a while.

What sort of future do you imagine after 2036? Will we colonize the solar system?

Keep in mind that not all humans were destroyed but we were all affected. There is an effort going into colonizing space because it is believed that the problems of overpopulation were a large cause of the war. Personally, my generation sees itself as having a duty to try and repair the mess our fathers handed to us. When we were young, most of us had a small taste of the world you live in now and our only dream is to clean it up and give it back to those still able to have children.

01-30-2001 06:21 AM

Sometimes I imagine what it would be like to approach the Wright Brothers in 1910,

272

before their first flight, and make the suggestion that in a mere thirty years, man would be on the brink of flying through the air at the speed of sound.

What tools would I be able to show them that would convince them? Would a picture of a jet airplane do it? Would complicated math and physics equations do it? Would it take a ride? Perhaps there will be a way to share the photos again but I don't expect it would convince anyone. I would only hope they would spark conversation and make the reality of time travel a little more personal.

Although I have no personal experience with non-mechanical time travel, I cannot discount it. Physics has a way of making the impossible a reality.

I'm not here to study anyone. My objective was in 1975 and the reason I'm here now is my family. I find my preconceptions of what I would encounter interesting. Being exposed to a society through its art, music, and advertising is one thing and experiencing it is another.

I'm not sure the physics of time travel is really that hard to grasp. Most of the working theory has been around on a large scale since 1970 and the technical breakthroughs are happening on your world line right now.

Technology is not gone in 2036 nor is it the private domain of "government" leaders. Computer printers just didn't work very well on 12 volts and many people just got used to doing things the old way. After the war, the main problem was distribution. Can anyone tell me how many companies in the United States still manufacture bicycle tires today? Anyone who still has a bike in 2008 will find out.

01-31-2001 02:14 AM

I'm not aware of any predictions I made or perhaps we do not agree on the definition. What anyone chooses to do based on something I might say will not affect me in the least. My goal is not to believed and I submit that your life would not be any better (and perhaps worse) if you did believe me. You placed "tests" before me that I must pass. Why? What do I have to gain by passing them?

The fact is there is nothing I can say or show you or let you drop hydrochloric acid on that will "make" you believe me and I really don't want that. It would be nice to discuss your view on religion, politics, physics, and the mechanical requirements of time travel but in order to engage in those types of conversations, I must apparently tell you who wins the hockey game next week. I'm just guessing that if were to write out the ten lines for Fermat's final proof you wouldn't be very impressed either.

(1) Industrialized mass production does not produce the uncountable tonnage of useless consumer items so gleefully absorbed by your society. I would estimate there are about 10 units like mine (C204) and twenty larger units (C206). The main difference is the sensitivity and number of the main Cesium clocks. I would estimate that some sort of public time travel will be common around 2045.

(2) I'm not aware of any other time traveler's "here" now. But if they are here, I'm sure they're pouring over sports history books so they can go back in time another week and start a friendly conversation on the web.

(3) The "Mad Cow" story here is yet to begin but don't worry, the fruited gelatin

deserts are safe.

(4) I'm glad to see it's so easy for to dismiss the Middle East. Yes, I suppose it is a no brainer but pretty soon it will be a "no arrmer" and a "no legger".

(Note from Lyn: remember folks, this was prior to Fukashima!

Ethics is an excellent topic of discussion and I hope we can move past a collective insistence of applying everything to this frame of reference.

01-31-2001 03:41 PM

The distortion unit reaches its target destination by using very sensitive gravity sensors and atomic clocks. The basic unit of calculation is the second. So yes, in a sense you do "dial in" in a date and the computer system controls the distortion field. At maximum power, the unit I have is capable of traveling about 10 years an hour.

Unfortunately, time travel is not an exact science. There is inherent error and chaos in the computers ability to make accurate calculations. Based on the current technology of the clocks and sensors, distortion units are only accurate to about 60 years or so. So no, in 2036, we are unable to travel back 1000 years due to the error rate in the system. The divergence between the world line of origin and the target world line would be too great. If one were to try and travel back that far, history would look nothing like what you would expect.

The unit has mass limits but the 204 is capable of transporting about three people and equipment. I don't think you would like 2036 very much.

The 5100 had a very simple and unique feature that IBM did not account for and decided it was not in their best interest to advertise (which in hindsight was not very smart). This accidental feature was thus removed from any future desktop computers. In order to take advantage of this feature, the 5100 I have now required a couple of special "tweaks" that had to be done by one of the software engineers in 1975. Anyone who is familiar with this feature and was told to keep their mouth shut about it will be able to tell you what it is.
Yes we still have toilet tissue and some people still suffer from extreme anal fixation.

I have noticed and gotten used to the act of verbal conflict as a cathartic entertainment. I don't totally understand it but I take no offense by it either. Perhaps we could just arm wrestle some day and still be able to have a pleasant conversation.

The year 2008 was a general date by which

time everyone will realize the world they thought they were living in was over. The civil war in the United States will start in 2004. I would describe it as having a Waco type event every month that steadily gets worse. The conflict will consume everyone in the US by 2012 and end in 2015 with a very short WWIII. (*Note from Lyn: Could he have been referring to the corrupt World Wide banking system?*)

The source of power for the C204 that allows it to distort and manipulate gravity comes from two micro-singularities that were created, captured and cleaned at a much larger and "circular" facility. The dual event horizons of each one and their mass is manipulated by injecting electrons onto the surface of their respective ergo spheres. The electricity comes from batteries. The breakthrough that will allow for this technology will occur within a year or so when CERN brings their larger facility online.

Perhaps it would have been clearer to state that the math has been around since 1970. I would urge you to examine the properties of Kerr black holes and Tipler cylinders. An actual working prototype was first tested in 2034. On my world line, time travel is not a public recreation but we are all aware that it exists. You may be disappointed to know that the ability to manipulate gravity is not the technical challenge that had to be overcome.

Miniaturizing the clocks and sensors, creating clever ways to vent x-rays and creating a computer system dependable enough to calculate the changes required to the field were the main challenges. There are no missing pieces, just missing energy levels and a few very interesting subatomic particles.

February 2001

02-01-2001 08:36 AM

Unfortunately, winners of historical sports betting events are not high on the priority list of people in 2036. As a thought experiment, if I did tell you who the winning horse was and you killed it before the final race, would that make me a liar or would it support my statement that our world lines are about 2% different? Do you know who won that race 30 years ago?

02-01-2001 11:28 AM

If you could change one thing about your government right now, what would it be?

The United States is still a representative republic in 2036 but it was touch and go for a while. After the war, the U.S. had divided into 5 general areas based on their economic and defensive strengths. Many people blamed the government organization for the war and the last Constitutional Congress was held in 2020 to officially scrap the Constitution and start over. Fortunately, this exercise in anger pointed out how hard it was to come up with anything better. It was decided the document wasn't at fault. As a result, there have been a few small changes to the Constitution and the executive branch but you would easily recognize it. The average citizen is more educated about the Constitution and aware of the rights and responsibilities it gives them. Federal power has been decentralized and the focus of daily politics is in the state senates. Federal law has also been streamlined but much harder to change or make additions to.

The people who sign my paycheck told me why we needed a 5100 and sent me off to get one. I was not in a position to make alternate suggestions. As I recall, isn't the Cray a rather large system? We need something portable. The 5100 isn't required for its reliability, its needed to translate between APL, UNIX and a few obscure IBM mainframe languages.

The fishing is great and you're more than welcome to join us but the "me" here is only three. I'll have to tell him your coming.

02-02-2001

Yes, I age and my hair and nails grow at normal rates. Please keep in mind that gravity distortion does involve some dilation effects but "jumping" between world lines are time-like trips, not space-like trips.
The air is about the same although I do smell and taste industrial odors here my parents cannot. The food in the future is grown and raised naturally inside the community structure. This is done primarily for safety reasons. I am amazed at the risks people here are willing to take with processed food. All of the food I eat here is grown and prepared by my family or myself.

Unfortunately, I do not know if we are acquainted on my world line. Yes, love is a challenge. What's harder is knowing you could go back and correct a mistake but at the cost of the "you" on that world line.

The questions about the president and space travel are reasonable but now we come to a conflict between physics and ethics. First, the ethics:

I have seen a television program about a man who is able to speak with the dead. When I watch the show, I am more afraid about the possibility that what he is doing is real not weather or not he is doing it. Since I will be leaving this world line in the coming year, I could easily tell you that the President lives or dies in the next four years. In fact, I could probably find some way to even charge you for it.
When the day comes for my "prediction" to be realized it will either happen or not. If it does happen, then your ability to judge your environment is crippled by your acceptance of me as a "knower of all things" and gifted with the ability to tell the future. If I am wrong, then everything I have said that might possibly have made you think about your world in a different way is suddenly discredited. I do not want either.

Although I do have personal reasons for

being here and speaking with you, the most I could hope for is that you recognize the possibility of time travel as a reality. You are able to change your world line for better or worse just as I am.

Although this will make me a far less interesting time traveler, these are the rules I personally try to hold to:

1. I will not disclose any information that will cause someone to personally gain by its knowledge. This means no stock or sports tips.

2. I will not disclose any detailed information that would allow someone to avoid death by probability. This means no earthquake or bomb information.

3. I will not disclose any information that may compromise any future actions by individual people or threaten their family and well-being. I will not disclose names or events associated with individuals.

Now for the physics: The grandfather paradox is impossible. In fact, all paradox is impossible. The Everett-Wheeler-Graham or multiple world theory is correct. All possible quantum states, events, possibilities, and outcomes are real, eventual, and occurring. The chances of everything happening some place at sometime in the super verse is 100%. (For all you scientists out there, if Schrödinger's cat had a time machine, he might not be in the box at all.)

Therefore, there is a world line where you are alive and another where you have gone back in time to kill your relative.

Differences between world lines are measured from the perspective of the time traveler in terms of divergence percentage. The higher the divergence, the more "un-like" your destination world line looks like compared to your world line of origin.

Therefore, any "prediction" I might make has a slight chance of being incorrect anyway and you now have the ability to act on it based on what I've said. Can you stop the war before it gets here? Sure. Will you do it? Probably not.

As far as space travel goes, no, we are not on Mars yet but we're trying very hard so we can avoid another "Hell's Kitchen" outcome from an overpopulated Earth

02-02-2001 10:09 AM

Perhaps it would be better if you just considered me a fraud. I really don't have a problem with that. If that were the case, could we then have discussions that you were comfortable with?

If I knew someone was going to die, and there was a chance I could stop it, I have a moral obligation to do something about it.

I can think of a couple of examples.

If the Egyptians knew the Red Sea was going to drown them, do you think they would have pursued Moses?

If you could go back in time to 1941 and tell the radar operators to take a second look at the radar screen on December 7th, would you? Before you say yes and accept that parade in your honor down Main Street, perhaps you should go forward in time and see if the U.S still had the motivation to make the A-bomb before Hitler did.

02-02-2001 10:53 AM

I fail to understand why my words generate so much conflict. I think it's far better for you to consider what I say as fantasy so there is no question of credibility. How is my credibility going to affect your life? I don't want you to believe me and it doesn't affect me in the least if you did.

I don't know any other way to tell you that I am unaware of what happens in the next week, especially in Hollywood. Just curious, that's a common question, why do you think I would know something about that?

Yes, it's very possible that what I say would spin your future off into a different direction. But since what I say is "bogus" that shouldn't be a problem.

On a philosophical level, the existence of multiple worlds implies a moral balance in the super verse. For every world line you perform a good action, there is a world line where you perform a bad action. There are no good and bad people, just good and bad decisions. We can only be responsible for what we do as individuals on the world line we are on now.

So take heart! Somewhere out there is a world line where I'm spilling all the beans on Hockey, the stock market and Hollywood and you're all off to Vegas and Wall Street making millions of dollars.

I do very much enjoy these conversations and I'm working on the other questions.

02-02-2001 11:41 AM

As far as the war goes, my best advice is to find at least 5 people within 100 miles of you that you trust with your life.

No, I haven't met any other time travelers here and although from my perspective that's highly unlikely, it's not impossible. No, I don't have any additional information concerning crop circles, ETs or UFOs. I find those subjects rather interesting myself and it's one of the reasons I was drawn to this web site.

Can you give us some personal stories of your past?

I was born in 1998 so I do share some childhood memories with all of you. I remember going to Disney World at Christmas and I remember going to the beach in Daytona. When the civil "conflict" started and got worse, people generally decided to either stay in the cities and lose most of their civil rights under the guise of security or leave the cities for more isolated and rural areas. Our home was searched once and the neighbor across the street was arrested for some unknown reason. That convinced my father to leave the city.

From the age of 8 to 12, we lived away from the cities and spent most of our time in a farm community with other families avoiding conflict with the federal police and National Guard. By that time, it was pretty clear that we were not going back to what we had and the division between the "cities" and the "country" was well defined. My father made a living by putting together 12-volt electrical systems and sailing "commodities" up and down the coast of Florida. I spent most of my time helping him.

Outright open fighting was common by then and I joined a shotgun infantry unit in 2011. I served with the "Fighting Diamondbacks" for about 4 years. (Hearing in my right ear isn't as good as I would like it). The civil war ended in 2015 when Russia attacked the U.S. cities (our enemy), China and Europe. As unusual and bad as my childhood might seem, I wouldn't trade it for anything.

Africa is not a pleasant place to be in 2036 although I would characterize it as recovering.

The music you enjoy now is quite popular and available it's just not produced in anywhere near the same amount. There is a revival of "local" and classical music. Many people have learned to play their own. I personally enjoy Big Band, some Classical and interesting lyrical pieces from the 1970s and 1980s.

If time travel is possible, what does that do to the religious? If I exist on many time lines, which one is really me?

This is an excellent question that causes a great deal of controversy. Since every possible outcome, event and possibility is happening and will happen, then all good and all evil balances out in the super verse.

After the reality of multiple worlds sank into our collective thought, the one basic change to all religious dogma is the concept that good and evil does not exist as an organized force in our lives nor can it be used as a useful way to judge what God may think of a situation.

Good and evil are personal experiences that

can only guide what we do as individuals and how we relate to others. This outlook also makes it impossible for me to judge any other person or event. We cannot see the entire universe as God sees it therefore we will never be Gods or be capable of judging anything outside of ourselves. My actions can only be judged as good and bad by me and my God.

There is also an area of thought that maintains all of the "yous" out there will make up the "you" that eventually returns to God. In that manner, it is frustrating to know that you are capable of and acting on all of the thoughts and ideas you have regardless of what the "you" here is doing. How good or evil do you think you can be?

There is even an idea (supported in physics apparently but I have a hard time with this one) that there is some sort of communication going on between all of the "yous" that are out there. Some people think that memories, intuition and conscious are actually attempts by one version of "you" to talk to another.

Yes, we still have football and you will easily recognize it in 2036.

I am posting here because I enjoy talking with people without having to hide who I am, it's safe for me and my family and I can gather historical information from you and the web.

Yes, I have superiors. However, from their perspective, I will only have been gone a split second. The only real risk I am taking is spending too long outside my main line and risking a probability error (dying, accident, etc.). Based on the physics of gravity displacement, I can't leave when I ever I want anyway and I do have some leeway into how I conduct my mission.

It is impossible for me to change any world line that I am not on. Nothing I do here will affect my home. The "60 year flux" is a limitation of my machine, not of physics. Jet planes can't fly into space, my machine is not reliable past a 60 + year destination.

You've been to Cedar Key? You must be aware I don't know any personal information about you. If I was going to guess, I would say it's important to avoid the rattlesnake or you might be "walking" back to shore. You may be happy to know that although the bridge is gone, Cedar Key is still there.

The computer is from 1975, Rochester MN.

Yes, the Pearl Harbor example relates to Y2K. Have you considered that I might already have accidentally screwed up your world line?

Potential Hazard of being a Time Traveler --- Always putting things off until you can "get back" to doing them.

What's even worse is having the ability to go back and help the "you" there and you won't listen to yourself.

02-05-2001 11:28 AM

Yes, there is a log and my report and debriefing will agree with it. I have no reason to lie about where I am and why I'm here.

There are no crimes against time. However, if I were to go on some sort of a murder spree and I admitted it when I returned, then I would be held responsible for the crime of murder.

I am no more able to affect your world line than you are. Yes, I could make the changes you suggested but you must remember that there are an infinite number of worldliness out there where I didn't and I don't. In fact, there's even a world line out there where you're the time traveler and I'm the one writing the question.

I think the war would be good for you and your society. I don't want to stop it.

Nothing I do here will affect any other world line. "My" original world line will always be the same and yes I can make changes here that would make this world line different from mine but so can you.

02-05-2001 02:09 PM

ATTENTION! The following is a secret message for all mindless robot slaves controlled by time travelers. If you are unable to read this, please disregard it. This is an emergency, can someone do me a favor and look up everything they can find on Maxwell's equations, relativity and the formulas for volts and amps. I left all my physics books back in the time machine. Remember, just because I control your mind doesn't mean I think you're an idiot. MESSAGE OVER.

Yes, there are significant advances in computer science. It's interesting how easy it is to be fixated on the "machine" itself without any curiosity about many of the support systems. This is not really big news but if / then is out and if /then /maybe is in.

Yes I am with the "me" on this world line and he is three now. I find that fact to be rather hard for people to get along with so I don't bring it up directly. This is the second time I have "met myself".

The "machine" is owned by the military. I am part of a unit stationed in Tampa Florida. The real energy is in creating the machine, not operating it. It's interesting you should bring up the divergence percentage. You may find it entertainingly interesting to know that the divergence can be more accurate with more sensitive gravity sensors and clocks. The C206 uses 6 Cesium clocks and an active method of timing the changes in the atom.

02-06-2001 06:35 AM

We time travel to solve problems. A great deal of effort is going into repairing the environment and infrastructure.

I think there is only one God. I also think that our soul may be a combination of all the collective thoughts and actions of the infinite "yous". If that is true, it becomes very difficult to define death until all worldliness come to an end.

(1) Who wins the (sporting event) for the next 20 years?

I don't know. Even if I did, you could stop the horse(s)anyway and make it untrue.

(2) Who will be elected president for the next 20 years?
Please see number 2.

(3) Will there still be (popular motorcycle) around when you are born?

I was born in 1998.

(4) Please tell me the price of gold for the last 20 years.

I don't understand the importance of this type of information. Please tell me your opinion why this is interesting and worth remembering for 30 years.

(5) **Will it still be lawful for me to own handguns?**

I thought owning a handgun was legal in the United States? Yes, being familiar with firearms (along with the other responsibilities of the Constitution) becomes an important part of people's lives in thirty years.

(6) **Is it possible for you to bump into yourself when you are time traveling?**

Yes that is possible and there are no limitations on interacting with them. I find it interesting that there is some sort of collective negativity with the idea of doing that. Could it be that we are not really that comfortable with ourselves and therefore we cannot imagine meeting, liking or helping another one of us on another world line?

(7) **Can I go with you on your journeys through time? Anybody else want to go?**

I could probably manage three people with me. However, I would have to dump a great deal of archival material to get you in. I'm not sure you would like 2036 very much.

02-06-2001 08:33 AM

Russia and China have always had a very strange relationship. Even the news I see now indicates that continued weapons deals to allies, border clashes and overpopulation will lead to hostilities. The West will become very unstable which gives China the confidence to "expand". I'm assuming you are all aware that China has millions of male soldiers right now that they know will never be able to find wives. The attack on Europe is in response to a unified European army that masses and moves East from Germany. Also, please be aware that from my viewpoint, Russia attacked my enemy who was in the U.S. cities. Yes, the U.S. did counter attack.

Based on what I know about the 5100 (IBM computer), it has a few very interesting and worthwhile properties that make it worthwhile for a time traveler to recover. Also, please keep in mind that civilization is recovering from a war. Yes, we do have the technology but many of the tools were lost.

As you are probably aware, UNIX will have

a timeout error in 2038 and many of the mainframe systems that ran a large part of the infrastructure were based on very old IBM computer code. The 5100 has the ability to easily translate between the old IBM code, APL, BASIC and (with a few tweaks in 1975) UNIX. This may seem insignificant but the fact that the 5100 is portable means I can easily take it back to 2036. I do expect they will create some sort of emulation system to use in multiple locations.

When I arrived, I approached my father and was easily able to prove to him who I was. I am currently with my parents and the "me" who is three. They are very aware of what I am doing, why I am here and when I will be leaving. It may interest you to know that my father still does not truly believe the machine works even after touching and seeing it. Yes, education is still taught in a classroom but the entire focus and system looks nothing like what you have now. Don't worry, you won't miss it.

I still don't buy your story.

I understand your viewpoint and I respect it. However, I am confused by a twist in the way the language is used. Another fellow who posted earlier was a bit upset over what I was saying because he thought I was soon going to ask for money by selling something. Since I don't have anything to sell I am curious why, "I...don't buy your story", is the natural way to say what you did. I am aware that it's off the cuff to say but wouldn't it be better to say I don't believe your story? Why is the other way so common?

02-08-2001 09:40

I would characterize world politics as two boxers who have just gone multiple rounds and they're both pretty beat up. I'm sure someone out there wants to kill us but no one is very organized yet. There is a great deal of fear over rogue groups coming across un-launched missile systems, 55-gallon drums of Anthrax or portable nuclear weapons.

Korea United?

I guess you could say that. Taiwan, Japan and Korea were all "forcefully annexed" before N Day.

I don't remember a great deal about media coverage during the civil conflicts. I would probably characterize it the same way you see coverage of Waco, Ruby Ridge and Elian Gonzalez.

From my viewpoint, yes, this is an alternate timeline. From your viewpoint, no.

I have thought again about how to revel information that would make me more believable but I always come across the following problems:

1. All of you become much less interesting as sheep. I can't talk to you if you're not skeptical.

2. Anything I say could be acted on beforehand and changed anyway.
3. All the really interesting information is months or years away and I'd be gone when it happens.

4. I find it morally wrong to assist someone with anything where they might gain and someone else would lose or die.

5. There's a slim chance your world line is just different enough my "prediction" won't happen.

6. I simply don't know.

Consider that you are a time traveler who goes back in time to the first week of February 1970 and you are confronted with the same problem. What do you remember right now about the second week of February 1970?

Naturally, the conflict in Vietnam and the Middle East come up but as someone has already stated here, "that's old news". I suppose I could predict the failure of Apollo 13 spacecraft but since time travel is ridiculous, I would be blamed for sabotage. I might even decide to tell you about an earthquake in Peru but then people that would have died by chance will now live and vice versa.
All I can think of is to make something up. So here goes: The space shuttle mission may or may not have a problem connecting the new lab to the space station.

How was that?

02-08-2001 01:18 PM

I enjoy the conversation and I will respond.

The "pattern" of exchange in the war will not be a surprise. Many people will perish as a result of starvation and disease. I would also submit that you already know if you're safe or not. The trick is to not turn off your fear when you'll need it the most.

Australia is sort of interesting in what is unknown. After the war, they were not very cooperative or friendly (can't blame them really). It is known they did repulse a Chinese invasion and most of their cities were hit. They have a trading relationship with the U.S. but I would characterize them as reclusive and ticked off.

When people use phrases like "See what I mean", "You're not hearing what I'm saying" or "Something smells fishy", they are indicating the primary sense they use to process information about a situation. I find it interesting that my credibility and the phrases that describe it hinge on economic terms and whether or not I have something to sell. I don't. I also don't know how to clarify my position any better so I would suggest that if what I say angers you, it might be best to just consider it fiction. Soon you'll get bored and I will leave in a few months. Either way, it won't be an issue.

The "enemy" that was attacked by Russia in

the U.S. was the forces of the government you live under right now.

I think you are just some guy who likes to mislead people.

(1) I do not seek followers to mislead. I seek safety, animosity (did he mean anonymity) and good conversation.

(2) To me it would seem obvious that we both have a very different perspective on what's important right now in 2001. I would think that's what makes our interaction interesting. Would I be anymore believable if I told you I had just stopped a horrible event and you won't hear about it because it didn't happen? Again, this is the second time it has come up and I am very curious. Why would you expect a time traveler to know or care about what happens in Hollywood or some individual companies profits? You seem to think I have tomorrow's paper in front of me. Is that what time travel means to you?

(3) I never said I was a scientist. If this is about economics somehow and you hope to "buy" my story, then what do I gain by "selling" it?

(5) You already know that cars are dangerous and planes crash. I'm very confident you are capable of killing yourselves without my intervention. Actually, my moral obligation has nothing to do with you, it's between me and God.

Deception? Exactly what standard do you use to measure the truth around you? I have seen other threads with amazing and potentially real experiences on them. Why am I more threatening?

I have no memory of meeting an older me as a three year old. The events between world lines are isolated and nothing I do here will affect my world line. Yes, my parents are alive in 2036 but they have no experience with a time traveling "me" in their 2001 either.

02-08-2001 06:59 PM

1. For all of you interested in coming back with me to 2036, perhaps we should discuss the trip. Please be aware, the displacement unit moves through time, not space. First, we will be driving the current vehicle (Chevy truck) with the displacement unit in it to Tampa Florida. From there, we will go back to my arrival date on this world

line. Then we will have to drive to Minnesota, sell the current vehicle, and get another one that would have been around in 1975.

We will then move the displacement unit (500 lbs. or so) into the new vehicle and go back to 1975. Once in 1975, we'll drive back to Tampa and make the final hop to 2036. If you'd like to stay in 1975, you're welcome to do that. It can also get quite hot and stuffy during the trip and you'll be subjected to a 1.5 to 2 G force the entire time. You'll also need some sort of a re-breather system or oxygen supply.

I have found that many misunderstandings and arguments are based on the differences in understanding over a single word. Two different cultures can have a drastic impact on the meaning of words like "proof", "trust" and "credibility". My frustration is in understanding the use of the words. I am trying to follow the logic of all of the comments.

After reading the questions, I want to paint a picture that may help with the general theme of our collective experience in 2036. The war had very profound affects on people and how they relate to each other. As individuals, almost everyone in 2036 is very familiar with death. We all have stories of loved ones that have died from disease, war or acts of inhumanity. Most of us have even taken part in dishing the same thing out to the other side. As a result, we have become far more compassionate to the ones we love but mush less forgiving to those who don't pull their weight. We are more accepting of other's differences in our community because we depend on them to survive. We are also more conservative with our resources and closer to God because for a period, life on Earth was Hell.

The other major difference is in the concept of good and evil. With multiple worlds come multiple decisions and outcomes. For every good act, there is an equal and possible bad act on another world line. Taken to the extreme, this must mean that in God's eyes, there is no total good and total bad in the super verse. It balances itself out to infinity. I believe we are judged on the decisions we make as individuals and the good/evil I see on my world line is an illusion that has no worth to God. My reaction to it is what's important to God. Although this may seem rather heartless, it does allow me to see past the evil that people do and acknowledge the core of potential goodness inside them.

2. I am currently based in Tamp Florida in Hillsborough County.

3. Yes I am free but I have obligations as a citizen. Everyone is required to have basic military training and provide a period of time for community service. It is very much like a type of mandatory National Guard service + Peace Corps.

4. Yes civil liberties are more important. You will feel the same after having your house searched a couple of times.

5. We do not live in a racist society but definitely a prejudice one. Everyone carries their own water.

6. Our government looks very much like yours. It is a representative republic.

7. My definition of a patriot is anyone who defends the Constitution against all enemies, foreign and domestic.

8. Religion has become far less centralized. Extremes tend to bring you closer to God so I would characterize religion and spirituality as a much more of a personal institution.

9. I do not know what happens when we die. However, I don't think it will be easy. There may be some very difficult decisions we have to make when we die that we are totally unprepared for.

10. It's difficult to define the most predominant religion. Christianity has fragmented into many groups and people with different beliefs speak mush more openly with others who have different outlooks and opinions. I would imagine there are large groups of Hindus, Buddhists and Muslims but I have not come across them personally.

21. No, Jesus did not come back. At least not that I'm aware of.
If or when Jesus returns, do you think he'll be a lamb or a lion?

12. No mission to Mars but we are working on it. There is a group working on the idea of gravity displacement to get into space but the calculations and error rate are very large obstacles to overcome.

13. We don't know what's on Mars yet.

14. We're not on the moon yet either. A great deal of technical effort is going into

cleaning the planet up.

15. I'm not aware of any "mind control" devices being used on you now. However, there are a great many "non lethal" weapon systems in development that turn out to be quite lethal.

Sometimes I watch your television programs that show SWAT teams using new non-lethal weapons. They usually start out with, "In the future, the army and police will fight its enemies with new weapons systems..." When they use the word "enemy", they're talking about YOU! You don't really think the Marines are going to jump out of helicopters overseas with sticky goop, pepper spray and seizure lights, do you?

16. I'm not familiar with the term Psionics. I'll look it up and see if we just have another word for it.

17. Yes, Europe is a mess.

18. Yes, I think the New World Order idea tried to establish itself. I would consider them the combination of the old U.S. federal system, Europe, Canada and Australia.

19. When you say "teleportation", I assume you mean public transportation. Yes, we have a basic high-speed system.

20. I don't know if the government created AIDS. I tend to think yes but controlling it escaped them as it mutated. You can't have conspiracy to limit the population with a disease without the cure.

21. Yes, homosexuals are in the army. You tend to look past the individual differences of people when their job is to protect your life.

22. Yes, there are "gay" people in 2036.

23. I am a white male.

24. Yes, there is an entertainment industry. Again, it is very decentralized. The technology to express yourself with video is so readily available that many people do it all by themselves or in small groups. Much of the distribution is over the web. I would compare it theater here.

25. I don't know what you mean by VR?

26. The average lifespan is about 60 but I

expect that will improve as we get things cleaned up.

27. Yes, paper money is still widely used.

28. The 21st century has awhile to go but the most significant person in my opinion is the farmer-general who led us to victory.

29. Even for you, Tesla technology can be found in appliances, motors and generators. I would have to say we have come a long way on Tesla's foundation but we don't have electricity rays that cause earthquakes.

30. Getting back to my exact world line of origin is impossible but it depends on how you define the correct world line. I can get close enough so neither I nor anyone there would know the difference. It relates to the classic example of cutting a distance in half to reach it. You can always get closer but never there. It also has a lot do with neighboring universes on Penrose diagrams, but that requires more math.

31. Yes, we have the death penalty. We also have public punishment.

32. The most common crime is theft.

33. I don't understand what you mean by "race mixing".

34. Cloning is not common. However, altering the sperm and egg cell are common. It is very difficult to have children in 2036 for environmental reasons.

35. It's hard to say what the exact racial make up of the U.S. is. I would say it's probably the same minus 130 million people or so.

36. No, there is no segregation.

37. The education system is very similar to yours but the organization and implementation is controlled on a community and county level. Various communities also specialize in certain disciplines.

38-39. My comments on equal rights and "women's roles" could be quite lengthy and controversial. I wish to discuss that separately. However, women do fight in the military just as the men do but their role on the community level is very conservative.

40. Having children is such a serious issue in 2036 that birth control is almost

unheard of. It is very difficult and precious to have and raise children. Again, this is an issue I will be happy to go into later.

41. Yes, there are same sex marriages but it's not very common. Again, people have lived through so much hardship that if what someone else is doing does not affect them or harm the community, it's not a big issue.

42. South America went relatively unharmed. However, there is still a great deal of internal conflict with conventional arms.

43. Yes, people use drugs that alter consciousness but they are not very common.

44. Yes, I know who the Beatles are. Old music and other entertainment is available on the web.

45. There is another Pope but I do not know his name. I do believe that "you're" UNIX will also have a problem in 2038. I don't think that's a secret but maybe someone should put a 5100 aside for thirty years or so.

John

02-09-2001 02:02 PM

Waco, Ruby Ridge and Elian exist in your news archives. Telling you about impending plane crashes or other disasters (provided I could give you exact dates and times) may save lives at one point but cause cascading changes that take others at a later point.

I enjoy the posts because it's very hard to remain safe and have a conversation. I have often seen the classic question, "why don't time travelers revel themselves'? That's easy, if you believe us, you'll drug us into oblivion and put us in a small dark cell while men in white lab coats take a Phillips head screw driver to the magnetic lock on the singularity drive housing.
I will examine the web site you mentioned. I believe the 5100 is unique in its ability to run assembler language on the 360-machine platform and still be portable. I'm not sure if that fact was ever made public so it's the best "proof" I have. I would like to examine the software you mentioned; perhaps I can further justify my side-trip.

Yes, EMP took out a great number of electronic devices. That's one of the reasons why we don't have reliable technology laying around. However, in the opening hours of N Day, the Russians did not launch any high altitude detonations. They knew we would most likely clean up after them so they wanted everyone outside the cities to be able to communicate. Most of the warheads that hit the cities came in threes and exploded close to the ground. The heavy EMP damage was isolated to those areas.

Actually, you will probably be quite impressed with our Internet. It's based on a series of independent, self-powered nodes that are mobile and can be put up anywhere. It looks a great deal like your current cell phone system.

02-10-2001 09:49 AM
Thank you for the information. I will visit the site and I expect I will have some questions.

Yes, there are a great many people who think that Revelations has come to pass. However, the new fear is that Christ has returned, he's not telling us and he's very angry. I am not familiar with the Fatima Prophesies but I am aware of the term. I was not sent here to change anything. Since it would not affect my home anyway, it would be rather pointless.

I am here because of my family and the interaction I had with them in 1975. I understand the question about the antichrist but I must admit I hadn't given it much thought. My initial reaction is to consider how the antichrist would affect my life if I could identify him/her. If your life became a sort of "Hell" anyway, would it matter if the antichrist was real?

I became aware of this bulletin board through my various archiving work through the web. We do enjoy "talk programs" but they are distributed over the web. I don't have additional information about UFOs and I find the subject quite interesting myself. I have often wondered if they aren't time travelers with very sophisticated machines. Based on what I know about the nature of time I find this difficult to believe but nothing is impossible.

Yes, handmade items and the skills to make them are quite prevalent and people take a great deal of pride in their work. People with the skills to make clothing, work with leather and wood or cultivate a garden have become quite valuable. Please do not think your questions are unimportant. I enjoy the conversation.

Fiji? I've noticed that when most Americans think about Canada in this time, they think about pine trees, chooks and back-bacon. It may interest you to know that most Canadians in 2036 are some of the most efficient, ruthless and dangerous people I know. God help Quebec.

Within limits, I'm confident my anonymity is secure. The various email and web accounts I have do not have a big sign that says "Time Machine This Way". In order to be at risk I would have to be believed. In my experience, evil may be powerful, but they aren't very bright.

You couldn't even answer one simple question about music?

Unfortunately, your question is rather difficult for me. Any personal experience I might have with your world line at a specific moment is limited to my viewpoint and age. I can't tell you what music is popular in the next few years because the "me" here is only three years old. In addition, how does someone decide what the most popular music trend is? It would seem rather subjective to me. Thinking about it now I sort of wonder what it would be like to be a teenager now and know the answer to that question. When I was thirteen I used to worry about which shotgun slug would take a door handle off and whether or not I had any dry socks. Current fallout estimates are accurate but a bit exaggerated.

Your example of what people would do based on something I might say is accurate. However, my concern would be the potential actions of a single person, not the masses. In 1963, if I suggested that someone should watch the book depository windows as the President went by, the actions of a single person might have a huge impact on history. You also asked if I've visited my "past" family. That's where I am now.

Once a micro-singularity is produced and placed in a magnetic field, what is the approximate strength of that field. How is the magnetic field produced? What would happen if that field failed?

These are excellent questions that I probably won't answer to your satisfaction but let me make a couple of points first. Although I am familiar with the general manufacture of the unit, I am not a physicist or a scientist. My background is in history and I had the correct profile to accomplish my given task in 1975. I doubt that most jet fighter pilots are aeronautical engineers and I'll bet most of you couldn't give me the formula for horsepower, yet you drive a car every day.

The second point I can illustrate with an interesting story. In the opening phases of WWI, one of the largest problems early fighter pilots had was how to shoot a machine gun through the propeller blades without chopping it to bits. The problem was very straightforward. When the engine was off, the blades took up only a small percentage of the total arch but with the engine on, there was seemingly no way to know when it was safe to shoot through them. This problem lasted for quite a while until a very smart person came up with a very simple solution. By using a small lobed cam on the propeller shaft that controlled a switch on the machine gun, it was possible to stop the gun from firing only when the cam turned off the switch at the point a bullet might hit the blade. It was so simple in fact the other side only caught on to the idea after one of the planes crashed and they were able to take it apart.

My point is; seemingly very complicated

problems often have very simple and ingenious solutions using technology that is already available. The distortion unit is not magic and no alien technology was required to make it work. If you could see it, the "smarts" that went into designing it will amaze you more than the technology. Heck, the really interesting technology is in the computer.

The magnetic field does not require the fantastic energies you might imagine. The field is "created" and captures the singularity inside a very large and powerful specially designed particle accelerator. If the magnetic system failed (which has numerous backups including a system that would remove it from this world line), the singularity would evaporate. Although it is smaller than an electron, it would still be quite undesirable.

<u>Why do you keep telling us about the war? How do you know that will even happen in our world line? Something may have already changed and it won't happen at all</u>

Yes, you are correct! However, I am not confident things are different enough for you to avoid the conflict. You may also consider the possibility that a world with no war is far less desirable in the long run. In response to your other point, your assumptions about causality are correct but my personal morality still comes into play. I won't deviate from my three rules because of the way I would feel about myself.

The singularities do not create Tipler cylinders they create the same physical environment without all the mass. The same math works for both. There are two singularities. Their mass and spin is altered in order to adjust the size of the ergo sphere and cause the event horizons to interact and create the gravity sinusoid. I have a basic home schooling education (k-12) and a bachelor's degree in history.
I will get to the remaining questions soon.

John

11 February 2001 18:43

Greetings everyone! I've missed you all too. As I'm sure (name from forum) can tell you, my email system is made up of quite a few addresses, locations and computers. Some of them work better than others and for a period of time, I have been unable (in my own way) to get to this site.

It appears I've come back at just the right moment. I see you're all asking yourselves questions about who I really am, what I want and why I'm here. Frankly, I really can't blame you but perhaps it's a bit easier now to understand why I not only didn't expect you to believe me and I didn't want you to. As far as my credibility, I thought we had reached some sort of happy medium and we could call each other friend.

I see however that may not be the case and I must admit I'm a bit disappointed. It seems that unless I follow your expectations of what a time traveler should be doing, thinking or feeling than I must be a fraud. Or is it just safer to keep telling yourself that?

I think it was a terrible idea for John to have told us anything about his story.

Where was this sentiment and concern for my safety as I was telling my "story"?

John was always telling us he has nothing to prove. Why come to a time travel forum and talk at all then?

Is it really that hard to believe that plain old human interaction has it's own merits?

Although John is creative, his declarations on quantum theories have already been proven to be wrong.

Which quantum theories are you talking about? Please forgive me but if I missed a question or if there was something else to go into I would have been happy to do so.

02-12-2001 08:54 PM

Why would a time traveler reveal himself? In John's mission, these actions have no purpose.
I still fail to see why this is a problem. If you can think of another way where I can interact with people, I would very much like to hear your suggestions. Why isn't the mere act of speaking with other people an end in itself? Personally, I find the unit interesting and I like talking about it. There must be something in your life like that.

John talks about a civil war being started between the Democrats and Republicans. Later, this escalates into WW3 and he mentions that it is a

good thing for us.

I don't believe I ever said the war was between Democrats and Republicans. If I am incorrect, please point that out. It doesn't exactly escalate as much as it opens the door for other aggression. Yes, I think the war would be good for society and I would be happy to debate that with you.

John tells us to buy a gun, find 5 friends, and get a bicycle. This conflicts with his comments about killing us all off.

If you could point out how this violates the three guidelines, I stated earlier, I would give it considerable thought. Perhaps you are correct and that was a mistake. Again, I don't believe I said I want to kill you. Personally, I think murder is wrong. You seem to be pretty good at that yourself.

John mentions that he has no idea what GINGER is but he mentions media hype and alludes to IT not amounting to much.

Again, I don't believe I said that either. My point in that conversation was about feeling manipulated by hype. Since they ARE asking you to buy something, I would be very interested in your lengthy and critical postings about ITS credibility. If you could post that link I would very much like to see it.

John mentions the war was between the cities of America, Russia, China, etc.

Hmm, are you familiar with the Russian partisan movement in WWII?

John says the civil war leads to the world war in 2015. The civil war lasts for ten years?

It's 2004. I apologize for a missed key (very observant - we all need good critics). Perhaps our definition of war is different. I would define it as a conflict where organized groups engage in maneuver and armed conflict. The first U.S. civil war lasted 4 years and the English civil war lasted 6. How long is too long?

John says the mechanical typewriter is a major machine again in the future and the Internet still exists and is widely used.

Actually, what I said was, ".you may be surprised at how many people use typewriters." I don't understand the conflict. What's wrong with typewriters? At least you don't have to plug them in.

Thanks for the patience. I'm getting to the other questions.

02-12-2001 11:51 PM

Have you ever been on any other time traveling missions?

Yes, but they were all training missions. There is a great deal of psychological profiling and testing and one of the training missions involved choosing a time in your life (within two years) where you wish you would have done something different and then going back to convince yourself to do it.

The idea is to become familiar with the possibility of meeting yourself, which can be rather difficult. It is quite odd to look at "yourself" and have a conversation. Since any "you" on another world line would not be a mirror image, you get a slightly distorted feeling while seeing yourself, let alone the concept of speaking to yourself.

Could you post a link to the images of your time machine?

I would be happy to do that.

If you have any pictures of the computer you were supposed to get?

I'm pretty sure there must be something on the web now about the IBM 5100.

I looked up the name 'Titor' and couldn't find any. Is your family not listed in the phone book?

I guess it depends when you look.

Can I contact you through any other means?

I've tried talking online before and found it quite enjoyable but I find the questions and comments come so rapidly it's hard to keep up.

What is the exact date that the war starts? If you can't remember, can you tell us the month?

I remember the exact date. When it comes,

it will not be a surprise.

What is the speed of the average computer in the future?

GHz is not a useful measurement. Computers are no longer measured by their speed as much as the number of variables (not calculations) they can handle per second.

Can you time travel whenever you want? How much does it cost? Can you make stops along the way back to 2036?

No, the unit doesn't belong to me. I can't make stops on the way home as it will throw the gravity measurements off and it would force me to backtrack. Also, I can't just leave and arrive at any place and time I want to. There are physical and technical limits to when and where I can go.

Have you ever been in your own future?

No, I haven't been to the future of my 2036. My profile qualified me for a trip to the past.

I understand your position and appreciate your supportive outlook. However, I have given some very detailed information that could be checked out. Please consider that our frame of reference is quite different and finding subjects to discuss and exchange information about could be a challenge. I'm just not up on music, Hollywood and sports in the year 2001. You could ask me all you want about 1975 but I suspect that's not very impressive.

Actually it does, I have never personally experienced this time as an adult and (name of popular singing star) was not a big topic of conversation when you're hiking through the swamp.

Everything I know about your time is from books, magazines and old videotapes that weren't destroyed in the war. If it makes you feel any better, I do struggle with not talking about football.

The animal Kingdom is alive and well. I'm sure it suffered but there fewer people infringing on animal's habitats now. Nuclear war is a very undesirable thing but it is not the end of the world. There are areas and cities we can't enter and the environment did suffer a great deal of damage but we are recovering. Isn't Hiroshima a thriving city today? The major physical affects include skin cancer, infertility, infection, etc. Almost everyone has some sort of physical remnant from the war.

I am aware the concept of the Rapture is related to Christian Prophecy but I am not familiar with the details. Yes, there are people I trust here and I would hate for any harassment or harm to come to them. I am aware of the Mayan Colander but in 2012, it was not something I was able to think about. When the time comes, I'm sure people will find the signs they are looking for that leads them to the end of time.

When I say fear will keep you alive I am talking about the natural instincts and premonitions that we all turn off when it's convenient. The same person who has five dead bolt locks on their door will think nothing about getting into a parking garage elevator with a total stranger. I think the fear of God is the fear of separation from God.

Some of you may wonder what a time traveler does with his day while he's posting on the web. I spend a great deal of time downloading information and storing it for my return. (Popular radio show host name) site is definitely on that list. I'll have to think about who the most remembered people are. Again, my viewpoint is quite skewed.

Who is the major super power in 2036?

It depends on how you define power. If you mean military, the world has developed into an odd balance. There are plenty of nuclear weapons left but if anyone uses them they, will be instantly erased from the planet by everyone else, regardless of the politics. We are very tired of war.

Are the two political parties still Republicans and Democrats?

There are no Republicans or Democrats to speak of. There are now over 10 major political parties.

How many States are there?

The states, as you know them, still exist but their political power has been combined with other states around them. There are now 5 major geopolitical areas in the United

States.

Which states are the worst to live in?
When you can't drink the water it's bad everywhere.

Were only cities along the Eastern US hit in the Nuclear War, or all over the country?

Cities and large military areas in the entire country.

Do companies like (popular software and Internet names) still exist?
No.

Does money look different than it does now?

No, money is pretty much the same. Unfortunately, I do not have any money with me because I wouldn't be able to use it here.

When you time travel you can never actually go back to the exact place you left from?

That is correct. In physical terms, I can never get back to the exact world line I left from.

Was a lot of the US 18-24 age group killed in the war because of a draft?

Yes.

Do you ever get time travelers from later time periods in 2036?

Not that I'm aware of but I can't say it isn't happening.

Are there any other companies that send out time travelers?

I work for the military. GE just makes the unit.
Does the government know there that time travel going on?

In 2036, yes.

Why were you selected to be a time traveler?

I am related to a key figure in the development of the IBM 5100.

How many time travelers are there on your team? How do they choose new

ones?

My unit has between 6 and 10. When I left, there were 7 others. Military service, physical fitness, history or technical background and special relationship to target contacts get you in the door.

Which countries are on our side in the war?

Don't have much to add there.

Which country gets the worst in the war?

Again, the entire world is affected. Even if you don't take a direct hit, dying crops and no water can ruin your day.

Why can't you answer questions about natural disasters? No one can change something now that would cause an earthquake or stop one.

I will not share information that would allow someone to avoid death by probability.

I think you're avoiding questions like that. It just supports the fact that you

are not who you say you are.

Again, what I think is important doesn't impress anyone and although I could point to various things I've said about other subjects the response is usually ho-hum tell us about music and sports. I suppose I could lie and make something up but what's the point?

You should have at least a basic understanding of physics.

I do but your questions were rather specific. I would love to talk physics and I'll be happy to walk through the operation of the unit.

By exposing yourself here, you've very likely broken several time travel regulations.

No, I haven't.

Don't you put yourself at risk by doing so?

Yes, that's potentially true but what I gain offsets that. Does that statement answer the question of why time travelers do not revel themselves?

You should know a lot more about the machines you're operating.

I do know very much about it. I'm just not willing to share it with everyone.

There are inconsistent in several aspects of your posts.

If you could point out the specific questions I will be happy to address them.

Is the (car company) truck better than the (car company) truck for time travel?

The vehicle must have a strong suspension.

How did you buy your truck? How could you title it if you are only 3 years old at this point? How did you pay for it?

Don't worry. Fortunately, these are things we're taught at time travel school in "how to get around in the sarcastic 20th century". I said I didn't have any money from 2036. I have plenty from here.

Does the truck have to be running when you are traveling through

time?

The vehicle must be standing still.

02-13-2001 10:51 AM

I just have one question. Do you recognize the name (name from forum)? I want to know because I fully intend to be a prominent figure within the next 20 years.

No, I can't say that I do. However, since I'm archiving all of this and will submit it with my report, it will eventually end up on our Internet.

If you want to leave a message to yourself in the future, please feel free to do so. I would make it broad enough so your name or some other important word will pop up in routine search engine that "you" might be using in 2036.

Also, be aware that the "you" in my 2036 will be unaware that the "you" here left a message at all.

02-13-2001 11:23 AM

My whole reason for asking the

unimportant questions is because they are unimportant.

If it's unimportant to you why would it be important to me?

02-13-2001 01:51 PM

John has been unable to explain time travel. I will explain it here.

I could be wrong but I don't recall being asked to "explain time travel". If you could point that out to me in the posts I would appreciate it.

So it's ritualized combat on the battlefield of differential geometry. It's a shame we couldn't have more constructive dialogue on the subject. Of course, I've left my physics book in the time machine so unless you snuck in a hidden landmine, I found your opening move very straightforward. However, you're too confident I won't be able to offer an explanation that I'm sure other physics fans will appreciate. I wonder if Einstein and Grossmann did it this way or maybe they just listened to each other and tried to build on what they thought would work. Oh well.

Debunk my calculations on time travel. You can't, because you are a fraud!

When I look up debunk, I see: "To expose or ridicule the falseness, sham or exaggerated claims." The reason I cannot debunk your calculations is because they are true. They are not false, based on speculative facts or exaggerated. They are just incomplete. What you really want me to do is finish your explanation or I will be an imposter.

There are, however, certain quantities that do remain constant. These constants are related to four-dimensional quantities known as metric tensors.

Actually, I don't think that's correct. Minkowski space-time (4-D) will not allow you to use Pythagoras' theorem to describe tensors because time needs to be expressed with the opposite sign. (please excuse my change of variable case).

$ds^2 = -c^2dt^2 + dx^2 + dy^2 + dz^2$
(where ds describes time-like and space-like trips).

The tensor we should be discussing is:
$ds^2 = -a^2dt^2 + w^2(df - wdt)^2 + (r^2/$

D)dr^2 + r^2dq^2

I hope I got the symbols right but you should be able to recognize this...right? nuts... the a, f, r, Delta and q didn't make the translation in this font.

The US government has its hands in everything from biological tests to secret planes that can fly out of Earth's atmosphere.

Care to share with me how you solved the overheating problem on your space plane?
I will get to and review the questions I missed. I apologize if my answers seemed flippant. There are many posts I want to respond to and I am unable to pay as much attention as I would like.

02-13-2001 11:32 PM

As for John Titor's corrections on space-time manipulation, he has completed it correctly. However, he is still an imposter.

Apparently, I have made the leap from "fraud" to imposter. At least that's a start and I respect my opponent on his polite yet quiet concession on the other thread. I wish to emphasize a point I tired to make earlier. Even though I answered the question correctly, it doesn't really prove one way or another if I'm a time traveler and you should not think otherwise. I might just be really quick at looking up things up on the web.

I suppose we could debate whether or not I'm a fraud all the way up to the point I leave your world line.

14 February 2001 02:55

Actually, it's sort of a good feeling knowing you're out there keeping me honest.

02-14-2001 07:25 AM
My Motive:

I've been in your time a bit longer than I had expected. My next opportunity to go home comes in the spring. For most of my adult life, I have read about, wondered and debated about this time. I value this opportunity to share experiences. If you absolutely believed I was a time traveler, with no skepticism whatsoever, then we would be unable to communicate. The focus of our attention would then always be on the machine. The experiences, opinions and reasons you do things are just as valid as mine and just as different.

I hope to return home with a better understanding of why you think and believe the way you do. Although I do understand the reasons for asking, I won't gain from any communication with you by spouting physics formulas and pop culture predictions. Please do not assume I am purposely avoiding questions. I am human, I get tired, and I forget things. Please, just remind me if I missed a question and I will get to it.

I do have one tip though. If you want me to go over your post in detail put, "Hey John, you're a big Jerk." at the end of the insightful and logical part, not the beginning. In fact, maybe you could just abbreviate it and put a number rating from 1-100 next to it so I know how strong you feel. Something like, HJYABJ (78). It would save space.

The Physics of Time Travel:

ACCELERATION = TIME DIALATION

As pointed out earlier, acceleration will produce time dilation. This can be observed by the "twins paradox". As one twin stays on Earth, the other twin in his accelerating spaceship experiences a slower passing of time. When he returns to Earth, he is noticeably younger than his twin who aged normally in Earth time. This type of "time travel" should have been proven already on this world line with atomic clock experiments. With sufficient power, this type of time travel will only provide practical displacement in a future direction. This type of time travel is also isolated to a single world line. You will not meet yourself.

GRAVITY = ACCELERATION

As Einstein pointed out with his STR, the effects of gravity and acceleration are the same. Therefore, you will experience the same time travel effects in the twin paradox by being close to a large gravity source. In the atomic clock experiments mentioned above, the reason there was a difference in time was not because the clock in the plane was moving, it was because the clock in the well was closer to the center of the Earth. Constant speed is not acceleration.

LARGE GRAVITY = STATIC BLACK HOLE

The next step is to find a large gravity source to use in your time machine. Static black holes provide this type of power. As one twin approaches the event horizon or edge of the black hole, the other twin will watch him as he appears to slow down. He will notice his twin's watch run slower until it stops at the event horizon. The twin moving toward the horizon will notice none of this and see his watch running just fine. Although possible, a trip into a static black hole will not take you to another world line and it's one-way. The force of gravity will crush you.

ROTATING BLACK HOLE = DONUT-SHAPED SINGULARITY

Fortunately, most black holes are not static. They spin. Spinning black holes are often referred to as Kerr black holes. A Kerr black hole has two interesting properties. One, they have two event horizons and two, the singularity is not a point, it looks more like a donut. These odd properties also have a pronounced affect on the black hole's gravity. There are vectors where you can approach the singularity without being crushed by gravity.

DONUT-SHAPED SINGULARITY =

PASSAGE INTO ALTERNATE WORLD LINE

Another other more interesting result of passing through a donut singularity is that you travel through time by passing into another universe or world line. Please see Penrose diagrams for Kerr Black holes or you can examine the calculations of Frank Tipler.

So now the problem becomes where do we find a donut-shaped singularity?

A PONDERING HAWKING = MICRO SINGULARITY

Steven Hawking proposed the existence of micro singularities that were created in the big bang. They were probably about the size of a proton and disappeared over the years due to an effect of radiation evaporation. (Yes, black holes do emit energy.)

HIGH ENERGY PHYSICS = ARTIFICIAL MICRO-SINGULARITY

When I first started posting online a few months ago, I said that major breakthroughs in particle physics were around your corner. Soon, CERN will bring their big machine on line and they will be smashing very fast and high-energy particles together. One of the more odd and potentially dangerous items produced from this increase in energy will be micro singularities a fraction of the size of an electron. (ARTIFICIAL MICRO SINGULARITY = LOCALIZED KERR FIELD

Through trial and error, and although they are quite heavy, hot and capable of putting out a great deal of energy (300 - 500 megawatts), it's discovered that these micro singularities can be electrified and captured. It is also interesting to note at this point that electrified singularities also have two event horizons. By spinning these various micro singularities, a localized Kerr field is created.

LOCALIZED KERR FIELD = TIPLER SINUSOID

By using two micro singularities in close proximity to each other, it is possible to create, manipulate, and alter the Kerr fields to create a Tipler gravity sinusoid. This field can be adjusted, rotated and moved in order to simulate the movement of mass through a donut-shaped singularity and into an alternate world line. Thus, safe time travel.

I will continue with the individual posts next. Thank you for your patience.

02-15-2001 12:07 PM

The following are personal rules I try to keep (unless of course they conflict with my secret agenda). I look forward to discussing any discrepancies you may find.

PERSONAL RULES FOR TEMPORAL DISCLOSURE:

(1). I will not disclose any information that will cause someone to personally gain by its knowledge.
(2). I will not disclose any detailed information that would allow someone to avoid death by probability.

(3). I will not disclose any information that may compromise any future actions by individual people or threaten their family and well-being.

Thank you for your persistence and patience. It would appear some of my more sarcastic comments are directed at you. They are not and I apologize.

Your future is not unchangeable? Harm may come to a person if I define them as someone who will do something in the future? However, in this case, I just don't know. I am not familiar with pop culture in 2001.

Are you avoiding inquiries for more personal reasons? Her logic is sound and you're avoiding discussing some of the things that people naturally find curious about other cultures.

I very much want to discuss our cultures but please help me understand how you won't be able to change something I tell you happened on my world line.
What kind of music is popular, what kind of recreation is enjoyable, what holidays are important?

I'm not sure if you wanted to discuss these or not. If yes, I will be happy to do that.

Do you feel some pleasure in breathing clean air and not having to check a radiation counter every few miles?

Yes I do. However there is a fear about

being here that I can only define as uncertainty. When I walk around in 2001, the air smells clean but I wonder if it really is. In 2036, there is no gray. The air is either clean or it will kill you. That feeling is very overwhelming when I eat here.

Are there people in this time period who accept you as one?

I have a very few precious relationships with people online who accept me as real or crazy and don't ask any questions. Much of my email flows through them. My parents are the only ones that have access to everything I could use to prove who I am.

How are you financing things?
I have taken very clever and reliable measures to go undetected. Yes, there are probably people like that but I am not in active conversation with them. My expenses are not that large. I spend a great deal of time now archiving.

I spotted few typing errors in John's comments so I will assume that he has had an average education.

You must be energized and anxious to improve your education system then. Please tell me what you plan to do.

Name any near future event that makes history.

You mean other than the mad cow pandemic, the breakthroughs in high-energy physics and the unknown functions of the 5100? I realize I've only been on this board for a few weeks but I assume you've read the other postings I've made about these issues months ago in order to be so definitive.

If you are older than 36 then there should be 2 of you here now. You would both have the same fingerprints and DNA. If you want to prove your case then meet with your younger self and get some evidence.

I am with my younger self. I don't have a case to prove and I wonder how many needles I would be on the receiving end for that one. With your superior education, I assume you already figured out that pretty soon someone might try that with a clone. Be careful what you take for definitive proof.

How long will you be here? When are

you going back?

My first opportunity to go home is this spring.

What are you taking back with you?

A lot of hard drives filled with books, archived web sites, pictures and audio files. I'm also taking back family items that were lost in the war.

Is propane still around in the future?

Yes but not very much of it comes from natural gas. Hydrogen is converted into propane because it's easier to handle.

I would hope the location I live in would be spared (the Hawaiian Islands).

My parents went to Hawaii on their Honeymoon. My dad told me a quick story about going to a fast food store and paying 6 or 7 dollars for a hamburger. I got an image in my head of a huge tanker filled with frozen hamburgers headed into the Pacific. Hawaii is very dependent on the mainland for food, isn't it?

Thank you for your kind words.

Logically he could easily question people of this time and get all sorts of info without revealing himself.

Mediums like the Internet offer unique opportunities for communication. When I return, I will be debriefed on my opinions about how people in 2001 will accept time travelers.

Why would he reveal himself if he has no stated agenda for doing so?

I'm not sure I exactly said that.

Did John come here to give somebody a push to invent the Time Machine?

I find this one the most interesting. What do you think would happen if the United States, China or Russia suddenly developed a time machine and the rest of the world found out about it?

John makes a verbal maneuver that turns a question back on the one who asked it so they think he actually answered the question; which he just avoided answering. He does this all

the time and I wanted to point that out before it happens again.

I am forced to admit I must rethink what I know about Mobius loops.

Does the last name (forum personal name) have any historical relevance?

You may leave a message to yourself if you wish.

What happens with Australia?

I believed I wrote about Australia a bit earlier.

What colloquial language is used in the future?

Many people use the Internet for communication and entertainment. I would say that affects our speech. We type very fast.

What exactly happens to the water?

Yes, radiation affected the water but that can always be distilled out. There are biological hazards that cannot. In addition, fresh water is hard to come by without talking to someone with a gun first.

Have you met your parents here? What do they think of you?

Yes. I am with them now. I would say it's a combination of fear and relief.

Does time have ends?

Yes. It is believed that all world lines end. It is also thought that parallel world lines that appear to be the same end at different times.

John, if you were really a time traveler, you'd be dead. The Earth, the solar system and the galaxy are all moving. If you did travel back through time, you'd materialize in 1970 where the Earth will be in 2036, which is the vacuum of space.

This is an excellent point and one I thought I went over a bit earlier. There is a gravity lock system that compensates for the local gravity outside of the Tipler sinusoid. This is the reason the unit is only accurate to about 60 years.

You are inconsistent about not

knowing anything about physics but you're willing to discuss the operation of your machine. Which is it?

I suppose I am thinking about the physics and the engineering as separate subjects. I apologize for the confusion and I will be happy to answer your questions more directly.

My questions were ignored. When I asked others, they were skirted.

Perhaps we could just start over again?

You stated elsewhere that Australia repulses a Chinese invasion. Does this mean Australian government side with your enemy?

There were deep divisions in Australia also. I would associate it more with a powder keg than a civil war.

Does intercontinental transportation still exist? Have you visited other countries?

Yes, but the market is much smaller. No, I have not been overseas.

02-15-2001 05:06 PM

You have said you will not participate in helping anyone avoid 'death by probability'. Yet many things you have said could have caused an individual to do or not do something that will now result dying, or escaping death.

It would help if you could give an example. If you are referring to the conflict and war in your future, I'm not sure I'm specific enough to help any individuals avoid anything. Suggesting there is a war coming is a bit different than saying avoid Washington DC at 3:45 AM on March 12, 2015.
There is no way for you to be sure there is no future world leader reading this and believing you.

Are you sure about that? Besides, I think you can have just as much impact as any "future leader".

I don't believe you when you say you're not trying to prove to us you're a real time traveler.

I submit there is no way for me to prove

anything on the Internet; therefore it makes no sense to desire it. What exactly do you think I could do to prove it to anyone? I am confused by your term "the show". Do you feel my only goal here is to entertain?

Again, I am not unable to make you do anything nor would I want that.

I learn a great deal about your culture from the words you write (like right now). What do you think my goals are?

This thread has been all about you developing your story.

I'm not sure I understand this. How would "my story" differ it was "developed"?
I find it hard to believe that a software tweak done to a 1975 machine would be enough to justify a time travel mission.

Ah, something we have in common. Yes, I felt that way too. However, my job was to go and get it and not debate why they wanted it. I am not a computer expert.

A great deal of the computer infrastructure you depend on is based on very old systems and code. One of the reasons I was sent to

1975 was because of the person I met there, not the technology.

There are more effective ways to accomplish what you claim in this regard.

Perhaps you would share them with me. You might be right and I could make your suggestions when I return.

If indeed you support the model of time that all possibilities occur in different universes then I cannot accept you.

I'm not sure I understand what you mean. If you believe in Multiple World Theory, Hawking was not the one who first thought of that. If you do, then I must be real if all possibilities exist. As I recall, Hawking felt that it was possible to build a machine but some sort of vacuum fluctuations would destroy it right before you tried to use it.

You claim that you have no desire to prove your story yet everything you have done has flown in the face of that.

I'm not sure that's true. In fact, I've tried to

point out on at least two occasions that anything I do (at the request of someone else) to support my claims can be found someplace else on your world line right now.

What is more, you will tell us no names, no locations, or any specifics as a result of your supposed ethics. If those were indeed the ethics, you were committed to and reasoned with, you would not be here now.

I am curious about this also. Do you think I should not interact with you for your safety or mine?

Time travel may be possible; you would not land on Earth. You would land in a vacuum of space. You have to take into account that the Earth, the solar system, and the galaxy are all moving.

Yes, this is a problem. It was solved by taking a "snapshot" of the local gravity around the unit before leaving a world line and incorporating it into the sinusoid during travel. The short answer is, you "stick" to the earth but this is only a useful explanation to understand it and it's not practical. Since the computer system is using a virtual reference, the calculations become flawed. Thus:

1. Based on the accuracy and timing of the "snapshots" the distortion units are limited to how long they can travel before becoming unstable.

2. We must leave and arrive in areas we have prior or future knowledge of in order to avoid massive objects (buildings, water, etc...)

3. The unit has a fail-safe system during travel that drops out in case of a unit shutdown or radical departure in gravity readings.
Answer this and I will believe you until the ends of the Earth.

Again, you should not offer this to anyone for any reason.

If John wants a way to prove anything then I am willing to look over his proofs. However, just because I were to look something over gives no more meaning to the rest of you than it would if one of you looked his stuff over, believed it then told me.

Exactly!

How would someone that came here with no money, all of the sudden have plenty of money?

The reason I don't have 2036 money is because it takes up weight, space and can be faked and I can't use it for anything. What type of expenses do you think a time traveler would have that I would need so much money for?

Simply inscribe your knowledge of any large-scale events over the next six months. Encrypt said text file. Give text file to a custodian.

The only problem is, now you have to trust the person who brings the information forward.

Do you believe that faith alone will get you to God? Do you believe in an organized force of evil that works against men's souls?

02-19-2001 11:03 AM

I believe that faith and good works will get one to God. I believe there is an organized force of evil that works against God's plan for men's souls.

362

It's all part of the plan.

Please don't think me so cynical. I would never insult or degrade someone's religious views. My next questions would be "what about knowledge?" I am a firm believer that faith (and good works) is not enough to get to God. There is a mystery we must solve first.

I vaguely remember you mentioned being religious. How would you answer your own questions?

Yes, I believe in organized evil. It would sure be easier to carry out an "evil" plan if no one beloved you existed. Just curious, can anyone tell me what "Satan" really means?

Had to answer those quickly. I have nothing but open-mindedness for religious conversation and I look forward to more. I'll get to the other questions soon.

Thanks.

02-19-2001 01:14 PM

ARTIFICIAL MICRO SINGULARITY = LOCALIZED KERR FIELD Is this a

prediction?

If you can take that paragraph and find a way to make a dollar from it than more power to you.

I never said you're an idiot, I said you are aggravating. Which is not really a personal problem.

Point taken. I apologize.

You only discuss things that you care about. Seems rather one-sided to me.

I was hoping it wouldn't be so confrontational. I don't see how words can harm either one of us.

You must believe that we both have interesting things to say to each other. Isn't that worth it all by itself?

What state do you live in now in 2001?

I am in Florida.

Is John Titor your real name?
Yes, John Titor is a real name.

What is your agenda?

If I tell you, it would just be an agenda, which I'm sure, is much less interesting.

Is this dialogue between people in these posts part of your agenda?

If I had a secret agenda, talking to people would not be part of it.

Does your younger self realize what you are?

Yes, he is aware that I exist but he doesn't know who I am.

Are you married?

No but I did have a chance to convince myself otherwise.

What rank were you in the Army?

It is the equivalent of Major.

Where did you go to basic training?

I wasn't fortunate enough to go to basic. We were fighting a war at the time.

What company were you with, and what year did you graduate?

The organization of the fighting unit I was in fell under the militia. We fought against the organized army.

Do you have any fears?

I fear people who want others to take action based on their own emotions and irrational fears.

What are the rules for interacting with people who aren't from your time; do you make friends easily?

Have common sense and get your job done. Yes, I have friends.

What would the government do to you if they found you?

I'm sure I would end up in one of their nice little padded cells while they poked at my machine with a screwdriver. What do you think they would do?

What would you do if someone investigated you? Aren't you worried that it's only a matter of time before

you are found?

I don't worry about that very much. No one believes me anyway. Right?

And you said; "I very much want to discuss our cultures but please help me understand how you won't be able to change something I tell you happened on my world line."

Suppose I told you the space shuttle would have a problem landing at Kennedy tomorrow because something goes wrong with the runway. If someone with the authority to do so hears that and makes the decision to land at Edwards (than) bingo, your future has changed from my past.

I appreciate you answering these questions for me John, and thank you for calling me your friend.

I'd really like to believe that.

Rest assured, if I went to the authorities and said, "I know a time traveler" I'd get laughed at. The skeptics will always over rule the "believers".

If that were not the case, I would not be posting at all.

The US government, Russian or anyone else would be in the same boat. They would grab and hold any time travelers.

Yes, that's what I think too. The irony is, I'm not sure the machine will really do anything for them and all I can give them is stock quotes and sports news. (Just a little humor.)

John, please don't take my comments about being a fraud personally. I only want the honest truth.
I don't. Truth is something we all want. Like opportunity, its something you have to be ready for to recognize it.

John, what would it take to get you to stay after spring and leave during the next opportunity?

My parents are much better at cards than I am. I fear they may not let me leave in such debt. If I had all the time I needed, I would spend much more time downloading and archiving.

I fear that another month or 2 may not be enough time as I would like.

I will continue to answer the posts as long as I can.

In your "world time" have things like "Sasquatch", "Loch Ness monster", and other reported unusual animals?

We have our swamp monsters too. In fact, I think it's interesting that we all respond to the unknown the same way regardless of our cultural experiences.

Is remote viewing used to gather military intelligence?
I am only aware of it from this world line. I don't know otherwise.

Are psychics still common?

Yes, psychics exist but I don't have any knowledge of their use by the military.

Is there a police force, as we know it? Have any of your leaders been jailed or impeached?

Yes, we have police but they are organized in smaller groups. Yes, we still have political and religious leaders who find it difficult to obey the law. I would submit to you that the law is only as good as the people's willingness to apply it evenly and swiftly.

Will you post more pictures or pages of the manual?

Yes, I am pondering posting more of the manual. I am also considering having my departure videotaped and yes, it will be free. My only concern is how it might affect the "me" on this world line.

Is there a Global Superpower state?

Being a superpower only makes you a target. There is an uneasy balance in the world now that everyone "probably" has nukes, chems or bios. We don't just bomb people for the hell of it anymore. Military power is based on the number of autonomous fighting men who are actually willing to fight.

I guess you are held accountable for changes that occur in your time period.

You're pretty much correct about your statement but actually, nothing I do here will affect my home. I hold myself accountable for any damage I do.

In my opinion, people "now" take clean water, electricity and their feeling of safety for granted. If they leave the city in search of fresh water, they first have to make the realization that fresh water is a problem. It's much easier to demand someone else owes you fresh water than it is to leave the lifestyle that made it bad in the first place.

If I were transposed to the 1920's or 30's, you would have a hard time keeping me off the streets.

I agree with you. The first time I walked into a "superstore" I cried. I'd never seen so much excess in one place at one time.

It is interesting that the photos posted were posted by an "anonymous time traveler".

Those are mine. They were posted by another person who I'm sure is reading this also and would be happy to confirm that.

I see there is a hand control unit with a screen on it. I assume this is the computer interface. What does the

display show you?

Yes, that is a remote unit. The unit itself gets hot and "unapproachable" during long travel and you're usually subjected to about 2 G's. It gets a little difficult to move around and the hand held unit sits next to you. The unit displays many things but time in transit, time to destination, VGL variance and unit temperature are the most common during travel.

Also, there are 16 buttons. I have to assume further that these are multifunction keys.

Yes, the menus are screen driven.
19 February 2001 16:01

What weapons did you use to protect yourself? Do you have a weapon now?

I used a shotgun in the war. Yes, I think it's prudent to be prepared for anything.
What is your favorite food?

Oranges, I love them.

Your enemy was in the cities. Was the President in 2005 also on the enemy side? How did you feel personally

about the President then?

The President or "leader" in 2005 I believe tried desperately to be the next Lincoln and hold the country together but many of their policies drove a larger wedge into the Bill of Rights. The President in 2009 was interested only in keeping his/her power base.

While you were in the 1970's what did you think of Nixon?

To tell you the truth, I was more fascinated by the different standards Presidents are held to.

Do you plan to do any more time traveling when you return to 2036?

I don't know if I will be chosen for another mission. If I do, most likely, I will fly as an advisor or historical consultant.

Do you have any doubts about this time travel experience and the computer mission that you are on?

Absolutely. I have to believe there is an easier way to do this.

What have you learned the most from the people of this board?

There is a fire of intelligence and expression out there in the people who are criticized for their open-mindedness. I fear it will only be uncovered in the rest of the population through conflict. I was under the impression that most people were sleeping in this period but I suppose that generalization is too broad for any time period.

Where did you learn the use of "Gosub" in computer programs?

Gosub / Return. In one form or another, isn't that a pretty standard function in most computer languages?

Where did you learn to write English? How much time did it take?

I'm not so sure I write perfect English. I do read a great deal more than I see most people doing here.

Was China your enemy?

Not my enemy? I never fought any Chinese but their ability to hit Western cities with missiles made a lot of people unhappy.

Do they have (old computer games) in 2036?

I am aware they are video games but I have not come across them.

Does China have a manned space program between 2001 and 2036?

I believe they are pretty close to putting a man in orbit. It shouldn't surprise you if they do that soon.

Do you have any problems with the number thirteen?
Not personally.

If the world lines are changed by choices why do we still meet people that coincide with our previous world line?

Your world line takes form as the choices unfold. People do not disappear because you must follow the same physical laws that hold you here. (i.e. Information cannot travel faster than light on a world line.)

In your opinion, would timelines be

better represented as an inflated balloon, or a layered Rubix cube?

Balloon in balloon in balloon. A Rubix cube as in the toy, right?

Do you have television in 2036?

Information does arrive on video but cathode ray tubes are out and crystal or plasma is in. Distribution is over the net, not broadcast.

Do you have flashlights in 2036? If so, what type of batteries does it use?

Yes, we have flashlights and we use similar batteries for most things. We do recharge a great deal. Have you ever seen those wind-up radios? They're pretty interesting I think.

How is electricity generated in 2036?

"Most" publicly generated power is through very efficient solar cells. On a local or household level, there is steam, hydro and inversion generation. There is a debate on using a singularity to generate power.

Does the sun look different in 2036?

When you can see it through the high level smoke and haze, not that I'm aware of.

What sort of clothing do you wear in 2036? Is there strong attention paid to what people wear in the group?

Clothing is much more functional. I'm not sure what group you're talking about.

Where were you born?

I was born in Florida.

02-19-2001 07:25 PM

It took you five days to answer those questions. Is that the best you could do?

How many days seem normal?

You said you went to basic training on the other forum which is how you got involved in time travel. Be consistent.

Your question referred to army basic training. My earlier comment referred to be chosen for this mission.

You answered like a politician. You stepped around the truth without actually lying but it's logical enough to keep your story going.

I'm not sure but it sure looks like your trying to say I'm being truthful within your expectations. However, if you're trying to hurt my feelings, comparing me to a politician will do it.

I guess you are held accountable for changes that occur in your time period.

My only concern is how it might affect the "me" on this world line.
Actually, this is a good question. If the "me" here goes on to have the same type of life and future work that I did, it may not look good on his resume that another "him" has left a videotape behind of his future mission to 1975.

So where do you stand? You're not being very clear and you always give a conflicting point to other questions you've had to answer. Why is it always about you?

If you look at my concern carefully, you can

see that it won't affect me at all. It affects him.

John says no one believes him. Is that right? He's in Florida and goes by the name of John Titor. Show him how much you believe him (when you find him).

I'm touched by your concern for my safety.

02-20-2001 05:06 AM

John had a (50/50) chance of telling us the future.

Yes, there was a 50/50 chance of that happening but the odds were easily one out two that it could have gone the other way.

02-20-2001 09:58 PM

Perhaps it goes without saying but I would urge everyone to listen to (popular late night radio show host) show tonight.

20 February 2001 12:01

Thanks for your reply. Can you please clarify some of these points?

It's what I live for although I can't help but feel you're not exactly asking these questions for their own sake.

Describe what you mean by functional. Do you wear light colors, dark colors, loose, tight fitting? What materials are used for the clothing?

Seems a bit obscure to have a discussion about. Do we wear radiation suits? No. Do we use pockets a lot, yes.

How does this smoke and haze allow for solar cells to be efficient?

Do you have any idea how efficient solar cells are today? Sometimes it's sunny and sometimes it rains. Sometimes large dust clouds high in the atmosphere pass overhead. I don't understand the point you want to make. The sun looks the same.

Where did you learn to write and how old were you? How did you learn to type so fast?

I was home schooled and I spend a lot of time typing.

"Gosub / Return. In one form or

another, isn't that a pretty standard function in most computer languages? No."

I beg to differ on this point. I may not be a computer programmer but I do know that going away to perform one function and returning to the original function is basic to all software. Perhaps it may be called something else in various languages.

Did you have a historical consultant to help you?

No, fairly easy job.

What do you call these batteries?
Good and bad, same as you.

"Rubix cube a toy? Where'd you get that idea?"

I've seen references to them in various science papers. If I'm mistaken, what are you referring to?

2-21-2001 08:26 AM

I'll look for it, thanks. I have a couple for you (book recommendations). The Nine Nations of North America and/or The

Physics of Immortality (anyone recognize this author).

Would you be able to travel by going to the year 2001, flying a plane to a destination, then go to the future from there?

You can only travel in time from a static position (at least with the unit I have). In order to do this, you must have knowledge of the local terrain and building structures. That's one of the basic protocols we do in any time period for possible, future travelers.

If each world line is separate from the others, then wouldn't the consequences of your actions now have no effect on your original world line?

Yes, that's correct.

If this is the case, why won't you tell us things that will give us knowledge?

I am not qualified to judge if you deserve it or not and I have no idea if you may be the next (for lack of a better reference) Hitler. However, if I were able to physically help you from a situation because I was there and I knew it was coming, I would help you.

My only guess is that you are not a time traveler, and don't want to make anyone do something stupid.

Yes, I am aware that is the obvious first answer but I would hope my moral and logical arguments at least make a dent in your thinking. If you were a time traveler, would you be comfortable giving out all that information after considering the possible consequences? (Provided you knew it). If I were you, I would be worried about what the next time traveler might do, even by mistake.

How can I leave a message for myself in the future if the things I do in this world line do not affect any other?

Just post it here. All this information will probably end up on the web in 2036. If you're alive then and you think ahead for some reason to do a search on yourself, you might see it. Of course the "you" there would have no memory of doing it.

Your prediction of (national politics)

pending disintegration, beginning in three short years, is impossible.

Have you see the documentary on Waco? Just for argument's sake, what do you think would happen if information were discovered that confirmed the worst accusations made against the law enforcement officers there? Would you hope nothing?

The idea of a farmer general leaving his fields to lead his country's troops to victory is an old one.

Yes, I often think about that when I see pictures of "my" farmer general in Omaha. It's a large bronze depiction holding a shotgun in one hand a copy of the Constitution in the other. He is looking up at the sky in defiance of God after his father was killed.

I guess there are only a few stories and all of them are just re-telling of the same ones in different settings.

Just like life.

02-21-2001 01:29 PM

If you take a look, I am answering every question in order unless I see something quick that I think needs a response.
Thanks.

02-21-2001 04:22 PM

How far from what size city is it the safest to be?

A 10 Kiloton nuclear weapon will vaporize metal for about ½ a mile and have a heat effect for about 3 miles. A 100 Megaton nuclear weapon will vaporize metal to 35 miles and have a heat effect to about 250 miles. I believe the largest nuclear weapon ever built and tested was about 60 megatons.

As I recall, the popular strategy toady is to strike targets with multiple numbers of smaller warheads. The 100 Kt to 1 Mt are the most popular. I believe there are about 150 - 200 major cities in the US and half as many military targets. Please correct me someone if I am grossly incorrect.

You suggest a bicycle. What about horseback?

Yes, horses are good if you can feed and

water them. Also, it's very hard to eat a bicycle.

Should we be stockpiling guns?

The answer to this is NO! You will draw a great deal of negative attention to yourself. I recommend becoming familiar with firearms. This means taking a safety course and learning to shoot and clean many different types. There will be plenty of guns around when you need them.

What kind of people will be the ones least trustworthy?

The people with the most to lose if the world changes, Camel through the eye of a needle?

Is the conflict racial?

Not at all. In fact, I would say it goes a long way toward erasing racial problems.

Does the civil war start in such a way that those willing will have time to remove themselves to safer locations.

Yes. You will be forced to ask yourself how many civil rights you will give up to feel safe.

Will you readily be able to identify the enemy?

They will be the ones arresting and holding people without due process.

Does living near a river take care of water problems?

There is an odd saying that might be appropriate here. Safe is anywhere a hungry person can't walk in three days. Water is important but you must consider that when people need it they will know where to get it. I would not plan on planting myself permanently next to a water source. Yes, distillation dose make water safe but the runoff is highly dangerous. Please remember that distillation is not boiling.

How are the five people within the 100 miles contacted?

The goal is to have a place to go other than your house and to be able to trust someone with your life. Foster those relationships now.

Do communications stay intact

during the war?

Main communication systems no, CB, sideband and non-repeating short-wave, yes.

Be mobile. Set aside the things you absolutely would need and can carry on your back. You will not be able to stay anywhere indefinitely even with provisions and firearms.

Will soldiers be asked to kill their countrymen?

I'm not positive but don't they sign a small piece of paper now asking them if they would have a problem with that?

02-21-2001 08:20 PM

In this experiment the traveler only goes 30 seconds into the past. It seems that 30 seconds before his experiment was to begin he saw himself appear in the lab. There would now be two travelers and two time machines. It would appear that its a time loop and an infinite number of duplicates see a duplicate self appear in the lab thirty seconds prior

to the start of the trip.

Yes, that's possible.

What is the mass of the duplicate time travelers and where did the mass come from?

The other mass comes from other world lines. I like to think of it as standing in a room with mirrors on the walls and the apparent "me" in the room next to mine steps into the room from his.

What is the result of the duplicates arriving simultaneously?

Psychological confusion and a few fist fights.

How long will it take for the loop to decay?

The chances of hitting the precise world line where all the other duplicates are arriving is almost zero. It's possible but increasingly less probable with each arriving duplicate. The divergence decays and the world line is "less available" for new "yous" to arrive on.

What happens if the experimenter

decides not to continue the experiment?

He can always leave the room on his own world line or put a desk full of books in the position where the time machine is arriving every 30 seconds. That will probably trip the VGL system and stop the time machines from arriving.

Do you remember any poetry after 2001 in your past?

A Soldier's Winter

The day before it wasn't snowing.

The trees are strangers, leering, disapproving in the ash of winter ..my world, my life, my wandering path.

I pray God's eyes may once again gaze upon me and remind me that I am still His child.
I only (think) I remember the first line but the last one I remember. It has quite a few more lines that I don't remember. It is rumored this was written first as a letter by a soldier. After he died it was added to and edited by others. In my opinion, it has become a symbol for the collective guilt my parents' generation feels for what became of

the world.

Is there a market for new books in 2036? Are most of the books available old or new?

Yes but there are no large commercial printing and distribution companies. Books and other forms of hard media are distributed on the web and printed or put on other media from local hubs.

As a rough judge of character, have you read J.R.R. Tolkien? Did you enjoy them?

My father read the Hobbit to me as a child. I was always afraid of the dark riders but perhaps I admired them too.

02-22-2001 10:08 AM

I will get to all the questions. I'm trying to comment on them in order. I'm posting the names before all the questions so if you feel I missed something just bring it up again.

I saw something last night that I want everyone's opinion on. It's concerning two television commercials advertising the same cellular phone product. The first commercial I didn't understand right away but the second was obvious.

In the first commercial, a man dressed in cold weather gear appears to be in a snowstorm. He's on a cellular phone saying goodbye to his family as if he was going to die in the storm. We then see he is standing in front of a snow machine at a ski resort area.

In the second commercial, another man dressed in cold weather gear is talking on a cellular phone. We see a young women inviting him to a romantic evening. He seems a bit stunned and excited, hangs up the phone, and runs off. We then see he has abandoned an unconscious person he was giving emergency medical treatment to.

What do you think of these commercials?

02-23-2001 11:19 AM

I assume you do a lot of reading. Do you read fiction?

I am a big fan of Conrad, Twain, London and any type of religious Apocrypha.

What brought you to this website?

When I decided to present or revel myself as a displacement driver, I had been watching similar boards for quite a while. I believe the only way to accept what I have to say, as being remotely possible requires an open mind able to suspend major portions of the belief and logic system. In his own strange way, even (forum name who consistently confronts John) falls into this category. I would much rather talk to him than a straight line, give me the equations physicist. I don't gain any insight that way.

And here is another character question. Have you seen George Lucas's Star Wars Trilogy?
Yes I have seen them. I like the first one the best and the "next one", in my opinion, isn't that great either. That's a heck of a battery those light sabers.

What happens to Bill Clinton?

I don't really know.

What happens to Bill Gates?

This I do know but I won't discuss.

02-24-2001 07:05 AM

Why are orphans an issue?

Problems with the environment still have lasting affects on all people, which cause the average life span to (be) lower. In addition, people are more susceptible to accidents. Family life and children are very highly valued and the community takes the responsibility for raising children if their parents die.

What is your biggest environmental issue?

Water. You need it for everything and there is very little left in the world that is positively safe to drink.

Does distillation remove radioactivity from water?

It removes the dust and dirt particles that are radioactive.

Do people still watch TV?

Yes but it isn't broadcast anymore.

Are there women on your travel team?

Not that I'm aware of but I would assume there are women who are either trained or are training for the same type of work. I don't know why there wouldn't be.

What is the status of women in 2036?

I understand the question but I have nothing to relate it to in 2036. The status of women is the same as men. Equality issues disappeared during the war.

Do women hold office?

Yes.

Do women work outside the home?

Women are not expected to stay home and be "barefoot and pregnant" if that's what you mean.

Do women get equal pay?

Yes.

Are women safe on the streets at night?

There is still crime but people do not attack

each other the way they do here.

What do women wear for the most part?

Clothing is more functional. Women wear very similar clothing as men when working or training. In our free time or with our family and friends, clothing is much more individualized. Long dresses, knitted sweaters and pants are still quite popular. You'll have to forgive me; I'm not very good at describing women's clothing.

Are the Amish alive and well?

Yes, I believe they are.
What is the birth rate?

I don't know the exact figures but having children is radically lower than it is now. It is the one thing I wish we had that you enjoy here.

Is there an unusual rate of birth defects?

Yes. Mostly stillborn.

You said that your culture was centered around Universities.

Weren't they wiped out in the war?

Not all major universities are in large cities.

Do you use cell phones?

Yes, we use a form of cell phone.

Do people eat red meat?

Yes but not as much as you do.

Do people drive cars?

Yes but they are not produced in nearly the same numbers or used the same way.

Are airlines in operation?

Yes but again, not nearly as many.

Do airlines travel internationally?

Yes, but most people don't get really want to go overseas.

Are people pressured into a Christian doctrine?

Not at all.

Do the police make drug busts?

No.

Are there jails? What kind of criminals are in them?

Yes there are jails. Mostly theft, fraud, rape and murder.

What kind of public punishment is there?

Hard labor, community service, banishment (you must move to another community), public execution.

Your justice system sounds like New England during religious persecution and intolerance.

How do you define intolerance? We don't really have the energy or desire to waste time being intolerant. If you produce and help the community than you can do pretty much think and do anything you want within the law.

Are there herbal medicines and alternative life styles?

Yes.

Is there personal freedom?

Yes, the same freedoms you enjoy under the Constitution.

Is there an IRS?

Yes, we pay taxes. Sounds like you don't enjoy keeping track of your personal income taxes. I don't think anyone does.

Is the same type of currency used? Is ours today good in 2010 or 2020?

Yes, we use money. That's a good question. I don't see why you couldn't use your current bills in the future.

Other than time travel how do most people get around the country? Do people travel much?

There is a high-speed train system, horseback, bike, walking. Automobiles are used mostly for sport and some transportation.

Are people suspicious of strangers or are all you people just one big happy

family?

There is still conflict and mistrust. Yes, I am suspicious of strangers. I think that's an instinct we are given to help us stay alive.

How do most people die during the war?

In this order: Starvation - Disease - Bullet Wounds - Radiation.

02-24-2001 04:49 PM

Its been pointed out to me that the links to the pictures are all down for some reason.

If anyone has a public site I can post them again, I will be happy to see that they get to you.

02-25-2001 01:36 AM

John, do they have anything to do with the "future" of time travel?

There are numerous people and organizations that contribute to the practical application of physical time travel. I think you would be surprised how much real work is being done right now.

Let's assume you video your departure and your family sends the tape to someone. How would this affect you in the future?

It wouldn't affect me on my home world line in the least. I would only be concerned how it would affect the "me" here. Of course, it may be a large part of my secret agenda and I have no choice but to do it anyway.

I doubt that anyone could determine that you actually time traveled, but it would be a good show.

I wonder what it would have been like to see a plane break the sound barrier before the jet engine was invented?

You said that there will be a big war. Can you at least tell us which cities will be nuked?

No I won't do that. However, I submit to you that when the moment comes it will be absolutely plain as day that you are unsafe in the cities. The millions people that stay will choose to stay. That's what comes as a surprise.

Are we traveling in space in 2036?

Not yet but they are working on it.

Has first contact with an alien race occurred?

Not that I'm aware of.

02-25-2001 07:31 AM

You say your machine has a 60-year limit. Is it possible to go back 60 years and then again another 60 years?

Yes, that is possible but the divergence grows exponentially as you move farther away from your world line of origin. I could make 50-year jumps to go back and see what the world looked like 2000 years ago but there is a strong chance it would look nothing like what I expect. There are larger distortion units that are more accurate and have a larger window.

Have the people in 2036 proved the "world line theory?" Is there any information you can share with is that proves it?

The Many Worlds theory seems to wrap up

very nicely into current string theory. Unfortunately, we have not solved string theory yet either but (n-10) seems to be the best working model we have in 2036. As you are probably aware, the "big equation" does not need the final solution in order to take advantage of the smaller parts that do work in the real world.

You say you don't want to effect anything by giving information but you could change this world line just by talking about the war, or anything for that matter.

I don't believe that knowing a possible future makes it happen. You are capable of changing your world line for the better right now. None of the things I have said will be a surprise. They were set in motion ten, twenty, even thirty years ago. Are you really surprised to find out that Iraq has nukes now or is that just BS to whip everyone up into accepting the next war?

02-25-2001 01:00 PM

John Titor, I would love to come along with you for a ride to the future.

I appreciate the offer but I'm not sure you would like the year 2036.

What is the next movement in music?

I appreciate your frustration and quite a few people have asked me questions like this. The expected answer is that I don't want to break my personal code of "time travel ethics". The real answer is, I just don't know. I was not prepared for the year 2001, I was prepared for 1975. I don't suppose it would be very impressive if I told you Disco would be big until 1980.

Am I getting this? You can load up all the people who want to in the back of your Chevy pickup?

Actually, the requests were rhetorical. No one is going back with me.

Thanks. One thing I do find interesting about time travel tech is the expectation that we can pretty much go anywhere at anytime. These systems are quite complicated and they do have limitations. Are you going to be around in 1975?

02-26-2001 09:16 AM

Better then being a fanatic; who believes in people so blindly.

fanatic (n.) A person possessed by an excessive zeal for an uncritical attachment to a cause or position.

Nuclear warfare doesn't seem to be the biggest form of fighting in the future.

Nuclear war will be very effective at destroying an enemy's economy and the people's will to fight.

I doubt nuclear warheads are going to be shot from each end of the globe.
I would caution against that. That's exactly what "they" want you to think while they continue to develop smaller and more accurate MIRV's. Have you ever seen a neutron bomb the size of a basketball?

John, you say one of the hardest things to do in 2036 is find clean water. You also say you only trust food you've grown. Is any of this a result of your experiences with biological warfare?

Yes and no. Yes, biological warfare and

accidents do cause a great deal of problems but the lack of a working infrastructure also hinders the continuation of the food manufacturing you depend on now.

Is there any more background information you can give?

Once I leave, I would not want any attention to come to my family here.

I'm guessing the date of your return to the future is April 19th.

That is a day to remember but I was thinking more along the lines of March 21.

The most likely leader of a war movement like the one you describe would most likely be a Farmer.

Throughout history, farmers have often been a target of oppression because they are absolutely necessary to civilization but too busy to defend themselves. If you push a farmer too far, they stop growing food and have nothing to do but hide in the woods and shoot back.

My asking you if the mention of CERN as a prediction was a genuine

question and had nothing to do with making money. Is this comment a hint about the future?

Please do not be offended by my "making a buck" remark. I say it with a wink to help other people form their questions. Yes, some very interesting things will be going on at CERN in the near future.

What is interesting are the type of questions being asked, and the apparent hostility that you're being subjected to for no reason.

Yes, I find that interesting too. Sometimes I wonder what people are really angry about and I have come to the conclusion that frustration is better directed at the messenger. But then again, that's history.

I'm sure even in 2036 there is a tendency to tease one's detractors if they make themselves available.

I have no intention of teasing anyone but I do grow tired of the same cycle over and over again. Eventually, the people who do not like me or what I have to say (real or not) will win. I will either leave or grow tired of answering the same questions.

If your most vociferous detractor continually puts on weak attacks; it takes attention away from the really challenging questions.

Again, please do not confuse my inability to answer the same questions over and over with a desire to make someone upset. I gain nothing by angering anyone or making them look foolish.

He knows what "make a buck" and "more power to you", and "off the cuff" mean.

I know my English isn't perfect but I blame my parents for most of the phrases I pick up (wink). It's different sometimes seeing them in print than hearing them. It took me quite a while to shake off "sock it to me baby". "Cool" seems to be the longest lived phrase I've heard so far and "peace" seems to be making a comeback.

What book written 1884 is required reading for all physics students? If he doesn't know this it lends more evidence to the fact that he is not a time traveler.

Well, I'm pretty sure it's not the Principia and it looks more like something to do with Maxwell but to make your point, just about anything can be looked up.

What if John makes a recording of himself and submit that to a voice stress analysis.

I've heard a tack in the shoe works. It throws off the baseline "no stress" readings. Also, if you speak slowly enough, you can beat those programs.

02-26-2001 07:01 PM

Is your sense of "timing" off in new time environments?

No, my timing isn't off. I do however find myself stopping in mid-stride and paying extra attention to my environment when I forget "when" I am. When I was a child, my mother would tell me stories about angels. She told me that angels found it hard to communicate with man because man could remember his past but couldn't see the future. Angels, acting as the eyes and servants of God, had no memory of the past but had infinite knowledge of the future. Although I am no angel, I often thought about that story after I left 2036. Besides that, I do get a lot of colds.

John, you were born about halfway into this cycle. Gen-Xers would be about 40-50 years old in 2036.

In my opinion, the Gen Xr's ended up in two categories. There were the ones who had learned to be independent by breaking away from tradition and societies expectations and the others who had no idea how to take care of themselves and just wanted the trains to run on time. The ones in the first group feel very guilty about "letting" the world go to Hell and the ones in the second group are dead.

02-27-2001 10:15 AM

You say you were in the militia fighting the US Army. I would think civilians would have no chance of successfully fighting the military.

You must realize that why people are fighting is more important that what they are fighting with. The conflict was not about taking and holding ground it was about order and rights. They were betting that people wanted security instead of freedom and they were wrong.

410

Is it a stalemate with the resistance/militia hiding out until the cities are wiped out allowing them to surface?

The cities were not isolated because of them; they were isolated because of us.

You cite the approximate number of cities and military bases intact before the nuclear attack. Are they all hit?

Nuclear weapons and guidance systems are less than perfect. Most targets receive more than one warhead but some of them were more accurate than others. I would estimate the overall accuracy was around 60 to 70 percent.

Three day's walk from where?

In my experience, a motivated starving person is only capable of walking about three days. The more distance you put between yourself and anyone who is likely to be hungry, the better.

Does anyone stay neutral during the war?

Some try to.

Does anyone lead a normal life during the civil war?

No.

You say the civil war lasts from 2004 to 2008 and then the short big war in 2015. What do the years from 2008 to 2015 look like? How long does WWIII last.

I'm not sure I said that exactly. By 2008, I would say the civil conflict is pretty much at everyone's doorstep. Western instability during the conflict leads to the attack in 2015. WWIII is very short with a longer period of mop up. (*Note from Lyn: future historians may very well refer to this – our time of unrest: banking instability, world conflict, exploitation of innocence, etc., as WWIII.*)

You mention Canadians but I don't think you mentioned the impact on that country.

There's not a great deal I know about Canada except to say they were pretty much in the same type of conflict. They did have the Dew Line you know.

You said the position of women is controversial and conservative in 2036. What does that mean?

It certainly isn't disrespectful. I apologize if it sounded that way. It's one of those areas I realize will be difficult to discuss because we may lack similar experiences. In 2036, there is not a desire to "have it all". With factors such as the difficulty in conceiving and the decentralization of production and industry, there is not an unrealistic scramble to have a "career" and a family. Out of necessity and circumstance, family life has become more traditional. However, there are many families where the wife / mother is the main breadwinner and the husband / father remains "home" with the family. The difference is in the method of decision-making. People do not have children (if they are able) unless they can devote the required resources to maintain it.

02-28-2001 01:30 AM

Can you post an answer to my question regarding the "30 second" scenario that I asked about a week or so ago?

If I didn't get to everything, please expand your question.

02-28-2001 06:17 AM

In order to assist in where I am in the questions, I will post the page and person I left off with. It would also help if you could do the same when asking when I will get to yours. Since my time is growing short, I will be unable to answer questions that have already been asked in some form or another and I will make a note when I come across one.

Unfortunately, it has also come to my attention the proposed email system for sending out the pictures is not working out. Apparently, people on the receiving end of the requesting email are starting to have problems with their computers. They suspect it's coming from the "asking" email. Before I leave, I do plan to send out a few more pages of the manual and a video of my departure. I'm sure a method of will be developed to do that.

02-28-2001 10:21 AM

Actually, (name of forum member) is quite quick to catch many possible discrepancies in what I've said over quite a few sites in the last few months. For example:

What type of money system do you have on your world line?

Its not very different than it is now. Yes, we have money and credit cards. However, like everything else, the monetary system is decentralized. Banking is based mostly around the community structure. There are no multinational banking or computerized economic systems. There are also no income taxes.

Is there an IRS and a need to keep a lot of receipts and paperwork?

Yes, we pay taxes. I had considered going into more detail about the tax system but I didn't have a great deal of time. Currently, I am watching my father go over all his taxes and he doesn't look like he's having a very good time. My comment referred to the collective misery I see around me during this time of year.

02-28-2001 11:45 AM

I do enjoy the questions and I appreciate the interest.

I suppose there is a difference between a thought experiment and the real world. It appears we have our virtual laboratories confused and I'm not sure I understood all the rules to your experiment and then it occurred to me that in your position, this is all just a thought experiment anyway. I will try to be more literal in my explanation.

I'm not positive but I don't see anything that indicates the time traveler would remain in the same spot once he arrives. 30 seconds is almost long enough to get coffee in your thought experiment. If that were true, and they all kept moving, than the experiment could go on for quite a while until the planet filled with time travelers. You also stated, "it would appear" as a time loop. If it only appears that way, than the natural divergence may stop the experiment when three or four time travelers arrive and the others end up on different worldliness.

<u>The problem is, what happens to the duplicates as they simultaneously arrive at virtually the exact same location.</u>

Again, you use the world virtually, which to me means not exactly the same spot. Under the laws of physics, I don't personally know what happens if it were on exactly the same spot but I do know it's possible. Under the operational limits of the distortion unit, as soon as the VGL sensors pick up an unexpected mass in the target world line, it would shut down and drop off in a world line where your experiment is not occurring.

Given that their masses will occupy the same space, what prevents a naked singularity from forming?

I see, now they are in the same space. I suppose that's a possibility. If so, than the as soon as the experiment started, a singularity would form under the infinite mass and swallow the planet. Perhaps they tried this on Cygnus?

It's a nice creative answer, but it didn't address the question. The question was is it a loop; will it decay and how long will it take to decay?

Well, I think it's a nice creative question too. Under your example, the "loop" would terminate as soon as the singularity forms and would be constantly fed by all the arriving time travelers.

02-28-2001 12:28 PM

After taking a quick shower and listening to the quiet hum of my archiving hard drives, I decided that (forum member name) and I may have made bad second impressions on each other. I find it ironic because it's people like (same name) that will actually solve those physical issues and make it possible for people like me to go back in time and argue with them. So I hope there are no hard feelings.

It also reminds me of a short story between a bicycle maker in the 1900s and a man who could fly faster than sound.

BICYCLE MAKER: Well Mr. Mach, if your plane can go faster than the speed of sound, how did you solve the compressibility problem that would tear your flimsy craft to pieces?

MR. MACH: First off, the aircraft are much more stable and made of metal instead of wood and fabric. Second, it is possible to pass the sound barrier by designing the wings and body to move the shock wave down the plane as you surpass the speed of sound.

BICYCLE MAKER: Really? Planes made of metal? Well, if your plane can fly faster than sound then why don't you just fly to the moon?

MR. MACH: It doesn't work that way. You need air to make the engine function.

BICYCLE MAKER: I see. Your plane can go faster than sound but needs air to function. That's convenient and it all sounds like a penny-book fantasy to me.

MR. MACH: Perhaps, perhaps not.

March 2001

03-01-2001 09:58 AM

I must preface the following with a bit of melodrama. I feel a bit unqualified to answer the next few questions for the following reason. The way you and I look at life and death and its relative value is radically different. As any other soldier can tell you, once you watch a man's arm come off from a bullet you just fired or have been close enough to feel someone's last breath on your face, it changes you. I can only describe it in two distinct waves. The first feeling is power. You won when it counted and survived. All the personal shortcomings and faults you've carried with you your whole life just melt away in a savage euphoria.

If there's time to think about it, the next wave comes shortly after and is underlined by overwhelming guilt. You just killed someone and now God might be ticked off. Fortunately, the second feeling goes away quickly when the shooting starts again and gets shorter and shorter after every battle. After all, why would God put you in this situation? The point is, I personally do not like going through that cycle and the decisions I make about life and death might not be the ones you would expect me to make.

How come it doesn't bother you that people may die through your inaction yet you find it "morally wrong" that you might affect lives by active involvement?

I'm not sure I said it didn't bother me, I only stated I won't interfere on purpose. Again I refer to a historical example. Before Pearl Harbor was attacked in 1941, a small group of US soldiers were experimenting with a new technology called RADAR. As the Japanese planes were flying toward the island, they actually picked them up in time to thwart the surprise attack.

Unfortunately, they were unfamiliar with the equipment and figured it wasn't working correctly. As you are aware, the ruthlessness of the Japanese "sneak" attack galvanized the US people into entering WWII.

As a time traveler it would be easy for me to take a short hike up that hill where the RADAR operators were and point out to them that indeed the equipment was working just fine and they should probably call it in. Assuming they believed me, it is arguable that my lone single action could start a chain of events that would allow the US to meet the Japanese planes and stop them from attacking the battleships. As a result, the US people would still be angry but not motivated to enter the war fully since the Japanese were not a perceived threat. Thus, you don't begin research on the atomic bomb until well after Hitler has already dropped a couple on London.

I could save thousands of men on the Arizona at the cost of millions in London. I just don't know how one life will affect another. However, if I were standing next to a soldier who was about to be shot by a passing Japanese plane, I would push him to safety. I realize this is inconsistent on a small scale but I am tired of watching people die in front of me. If there is a price to pay for that than so be it.

Isn't it just as wrong to affect lives through inaction as it is through action?

Why are you concerned about what I might do to corrupt your world line when you have no problem letting other people do it around you every day? Do you blame yourself for not taking any action right now to "save" people living on your streets or suffering from poverty and disease? Besides, how exactly would you propose I decide who to tell and who not to tell? (provided I knew anything at all).

Your immediate decision, in itself, is its own authority.

What God judges about my decision is the only authority. Again, all the things you claim I can do you are capable of also.

If good and evil achieve a balance in

the larger picture, the question of you being required to decide who lives or dies is moot.

It's not moot to me. To tell you the truth, I'm afraid. I don't want the responsibility of being expected to know who lives and who dies. I know it would change me for the worse. Besides, how can you be sure my "inaction" now isn't a result of something I've already screwed up and I'm trying to fix it? Javier might be right after all. Thanks for the good questions.

03-01-2001 11:45 PM

Are the singularities in your machine supposed to be offsetting the light cones of particles within it's sphere of influence allowing the world lines of these particles to appear to loop form the perspective of particles outside the effected area?

No, that's not how it works. The singularities are used to manipulate gravity around the observer. The singularities do not interact with any matter except the electrons that are injected onto its event horizon. The hazardous areas of gravity are quite small and exist only around the inner singularity ring and another area created in the gravity sinusoid outside the vehicle.

I don't mean to insult you by spelling out the obvious but that seems the most obvious need of a singularity.

No insult taken. I would imagine we both agree that standing behind an operating jet engine is an unhealthy thing to do also.

If so, how can you account for generating a gravity well deep enough to create a disparity between light cones without sucking the planet through the eye of a needle?

The gravity well created by the singularities is not that large. The portion of the field that is felt by the operator is about the equivalent of 2 Gs. I would urge you to examine a Penrose diagram for a Kerr black hole. As you are probably aware, the singularity is donut shaped and exhibits two event horizons. The singularities are used to "simulate" a path through the center of one of these singularities, which is what takes the observer to an alternate world line. Earlier in the thread, I did go into this in a bit more detail.

Also, this requires motion through

the space immediately influenced by the mass, yet you claim travel is accomplished while the traveler is stationary.

The unit must be stationary during operation due to the sensitivity of the gravity sensors. Any motion with an acceleration component would throw the gravity measurement from the singularities off.

03-02-2001 08:01 AM

Are you posting on other boards without revealing yourself as a time traveler?

Not really, there are a number of science rooms and other chats I do visit and just sit and watch. I have discovered that people who frequent this board and some others have the most open and creative minds. I realize no one actually believes me but they are still able to

Are you in contact with time travelers in 2036? What percentage of people accept time travel in 2036?

The general public was informed about time

travel around 2034. Yes, I have had conversations with other time travelers on my home world line. Your insight on the public is more or less correct. I would say 60% of the people realize what it is and the possible implications, 20% of the people don't care, 10% don't believe it and another 10% see it as something that should be banned and stopped.

Imagine if a time traveler from the future came to this time period and told us the secret of time travel.

Yes, imagine that. Do you think that would be a good thing or a bad thing?
About the commercials, they are ignorant.

Back in the 50s and 60s, television commercials were straightforward. Usually along the lines of, "Cheese! It's good! Buy cheese." In the 70s, there was more identification with a producer or trademark but the commercials were still pretty easy to understand. "Buy this beer, it tastes great!" Today, I have no idea what some commercials are advertising until they show the logo at the end. Do you find this more effective? Only recently have I seen this move toward dark humor. I've never seen anything like it before, even in "your" archives in 2036.

Is spiritual awakening a difficult process?

I believe spiritual awakening is difficult. Why? I think God wants us back but the road we have to haul is no picnic. Maybe he's a little angry for some reason.

I think the world is seductively clever in its presentation. "It" wants us to stay here and it distracts us from God by creating want, greed and four or five other motivations. Our goal should be to; yes, have faith and do good deeds but also look past that and have the wisdom and knowledge to realize that this place, this world, this universe is not really our home. The question I ask myself is not can I get to God, it's, am I prepared for what will it be like when I get there.

There are plenty of great mysteries, but if your goal is to get to God', it is not necessary to solve them.

I mean mysteries not of this world. For example: I suspect that the final thing we will have to give up to get to God is our free will. Do you think many people will be standing at the pearly gates saying "yes" to that one if they had a choice to come back here?

What do you see for the future of time travel on your world line?

That's a good question. I am hopeful that one day when we get the planet cleaned up it will be a nice place to live on again and no one will want to leave it. On the other hand, if time travel were commonplace right now, I think a great many people would leave and perhaps never return. There is also a suggestion that time travel might make an interesting punishment. However, I don't think we have the right to force criminals on unsuspecting world lines and sending them to the Stone Age might be a bit much.

I thought I met another time traveler who said they were from 2036 once.

Although not impossible, I doubt there is another time traveler here from 2036. I have been chatting on and off for quite a while and in other chat rooms. I have also seen and heard about other people who have taken a creative license with some of the things I've said and posted. It might have been me but I've never heard of the

DNE.

I find this fascinating. Has California had the big earthquake?

The big one? As you are experiencing now, there are earthquakes, storms and other unfortunate surprises from Mother Nature that have impacts on your society and future history. That is one reason I won't go into detail. However, don't worry too much about major portions of coastline slipping under water.

03-05-2001 02:02 AM

After he offered to take people back with him, do most of you believe John is sincere in everything he says?

Would you do me a favor and point out exactly where I made that offer? I do recall a few people asking what it would be like but I don't believe I ever offered to take people back.

03-05-2001 11:23 AM

What music do 20 year olds listen to?

People listen to all types of music. A great

deal of it is available over the web. I would also add that people spend much more time making their own music.

What's the future of cloning?

Cloning full people has been determined to be medically and ethically unsound. We do have research and progress in cloning body parts and creating more viable sperm and egg production.

Any more on Bill Gates?

Not really. Just curious, why is he of such interest?

Do people wear chips yet?

No. People value their personal independence and ability to take care of themselves.

Are you a marked man?

Not that I'm aware of.

I would still like to know what population makes a city big.

Cities become targets because of their military and economic value. Any large area supported by a civil infrastructure is likely to be on that list.

Your time sounds grim. Are you tempted to finish your mission and then retire in the 1970's?

Not at all. I'm anxious to get home.

What did you think about those commercials?

I think those commercials capitalize on other people's misery and misfortune in an attempt to sell a product. I can understand coming up with an idea like but what confuses me is how does it get past that stage? How do people sit in a room around a large conference table and agree that leaving a critically injured person lying in the snow is funny and will sell cell phones?

03-05-2001 12:24 PM

Although your post wasn't truly an invitation to join you when you travel back, I can see how it could have been taken that way.

I'm sure many things by many people are taken in many ways. I find this an

431

interesting point because I think its important to have implied agreements on words and meanings before you can talk with another person. If it wasn't an invitation (by your own acknowledgement), am I responsible for what people think? If I am, how would you propose that I double-check that? Are there really that many people out there upset about this?

I'm still waiting for that public apology.

I publicly apologize for confusing anyone who is packing their old bell-bottoms and shawls for a trip back to 1975.
Just curious can you think of anything I could do to prove to you that I'm not a time traveler?

03-05-2001 02:50 PM

Is reverse speech used in the future?

I'm not very familiar with reverse speech but what I saw on the web leads me to conclude it's a bit objective. I'm not sure if its been proven scientifically to be very accurate. Are you aware of any research that shows that?

What about earthquakes in California or Nevada?

No I can't. Besides, I see others predicting earthquakes and very few people pay any attention to them.

Thanks for starting this topic and sharing your time with us. We are really enjoying it!

I appreciate that a great deal. Your future will be fine.

Warping time and space takes lots of energy.

Yes it does. A nuclear aircraft carrier and a space shuttle main engine also take a great deal of energy.

Hawking believes it's possible to build a time machine but a mysterious energy will destroy it if anyone tries to use it. In my opinion, manipulating gravity is not the hard part of time travel. Also, with great power comes great responsibility. If man has a limitation, that's it.

What if something happens to your device and you couldn't leave. Would you do anything different?

No, I wouldn't do anything different if my machine broke. I would still be a stranger and a guest here. My opinions and "announcements" would also be the same as anyone else's. I may however offer advice to my younger self.

I still have some questions you have not responded to.

If I missed something feel free to bring it up again.

Why aren't you traveling and telling us about your trip to the pyramids?
Yes, that would be fascinating but the unit I have is unable to go back that far accurately.

Instead of talking about specifics, you should be bored from your awareness to them.

Not at all, I find the subject fascinating. There are two real issues I hope people think about when I'm gone. One, how will you react when another time traveler shows up and two, how are we going to handle the responsibility of time travel when its invented.

Are the Great Pyramids standing in 2036?

Yes, although one of them was severely damaged.

If you wish to experience society as it was, admitting yourself to be a time traveler is counter-productive.

Yes, if I was here for that purpose and if you believed me I suppose that would be an issue.

How's communication around the world in 2036. Do you still have literature widely available?

Yes, books and other literature are available but most of the distribution is via the net.

What's the latest book you've read that you were only able to hear about in your own time?

The latest book I read was the autobiography of the Red Baron compiled from letters to his mother. Yes, I was aware of it in my own time but finding an original copy there was almost impossible.

Is new literature also so available?

Yes.

Is the English language segmenting? Is tourism still strong and thriving in 2036?

I would say the English language is pretty much the same as it is now. There are differences in slang and figures of speech but it's nothing you couldn't pick up. Yes, I suppose we do have "tourism".

Can you tell me when the police stop busting people for weed?

It happens about the same time they stop coming to your house when you dial 911.

Do they start pushing for legalization earlier than the war?

It's not really an issue of the government letting you do something; It's more like they have other things to worry about. Don't you feel you're capable of taking care of yourself? If you want to take mood-altering drugs, why should my opinions stop you from that? They don't stop you from taking alcohol, tobacco or fast food.

In any case, it also lets Darwin take over. One of the reasons drug abuse isn't a major problem in 2036 is because no one wants to die from it and everyone else who did is dead.

03-07-2001 04:48 PM

In the 18th Century, a pantomime Clown was known as John Trot, Clodpole and Clodpate.

Unfortunately, I'm not that well read. But it does look like an interesting story and I will be sure to pick it up if it sparks a connection between it and me. Madam I'm Adam...that's the only one I know.

03-08-2001 03:39 AM

I believe I have exposed John with the anagram in his name.

I find this interesting because it gives me a very tempting easy out. I could now rest assured that someone had "figured me out" and I can relax before I leave.

However, I am not (other forum member)

and this name and TTO are the only names I've used online. After looking at my name here, have you considered its origin from another word-play standpoint? For example, TITOR could equal Time-Travel-OR.

After looking at your name (forum member name), I can pull out "MEET ME TAR BABY" which I'm assuming is a reference to the Song of the South. In that case, are you telling us in a secret way that you are trying to distract us by fooling us in the briar patch?

I would not insult your collective intelligence by leaving a hook out there for you to discover while I was making sport of you. Whether I'm a time traveler or not, I think we've spoken about many important things I would not want to diminish.

08 March 2001 11:23

<u>The fact that he said he could not complete a 0 divergent trip meant that he could not return to his timeline and a mission into the past to help his people is logically flawed.</u>

I thought we went over that to your

satisfaction? Doesn't everyone know after looking at a Penrose for a Kerr singularity that you have to travel faster than light to get to the "exact" same world line?

I can see your not amused that we would be confused as the same person. I did find it flattering. I think you find some of the physics questions we're dealing with on other sites quite interesting...perhaps even convincing.

08 March 2001 17:37

Your argument was not sufficient. You suggested that an alternate you may return to your world line. It just does not seem logical.

As you are aware, approaching a rotating singularity can be done quite easily without experiencing the negative side effects of a massive gravitational field and it's very possible to "pass" through the center of the ring. Besides, if you did need a naked singularity, all you would have to do is increase the rotation or electric charge so the inner even horizon equals the diameter of the outer event horizon.

I realize my posts here have become

tiresome and my "story" is old so If I don't post it's not because I don't enjoy the physics debate. If you're interested, I will be posting more pages from the manual and a cut-a-way drawing of the distortion unit.

03-09-2001 07:11 AM

Greetings everyone. I do plan to get to the questions soon. I have been quite busy lately so I apologize for being a bit slow.

In my travels over your web, I came across this section of a speculative news article. I would urge you all to take a good hard look at this idea and consider the possibility that it is true. And no, I did not make this up nor am I trying to tell you something in a left-handed way.

03-09-2001 08:53 AM

(Forum member name) mentioned a few posts ago that he thought possibly the (other forum member name) persona might be a "sophisticated dialogue" construct of our time-traveler for self-dialogue.

Had you considered the possibility that (other forum member name) is the one who made me up?

Dear Fellow Time Travelers:

In about 30 days, I will be leaving this world line to return home to 2036. I first want to say thank you for the wonderful conversation and insight into your society. I have learned a great deal and my opinion on quite a few things has changed dramatically.

I will finish the questions that have been posted on this site up to this date. Unfortunately, I must now spend my spare time preparing to leave and I will not be on the computer very much. I do however want to repeat my offer and add a slight twist.

After going over my flight plan home, I have discovered my VGL holdover period is a bit longer than I expected. I will be spending at least three weeks in April of 1998 as I make my way back to 1975. Therefore, I not only offer you the chance to leave a message to yourself in 2036 but I offer you the chance to leave yourself a message in 1998. I will take any compiled messages and email addressees you provide and send them on the net when I get to 1998.

Granted, this will not affect you on your

world line now but you make take some comfort that another "you" on another world line has the advantage of knowing something you wish you knew three years ago. Based on the earlier questions I've seen, I've decided a day-to-day record of the Dow a day in advance should convince you that the messages are real in 1998.

In addition, I am hopeful a series of photocopies and photographs will be available for you that may give you more insight into the technology of the distortion unit. I will let you know the address of the site when it is available. I also plan to have my parents videotape my departure. If they succeed, it will also be posted after I leave.

I look forward to these last few weeks with my family and I will check in periodically to check this site.

Live in Peace 2001,
John

10 March 2001 22:35

Dear Fellow Time Travelers:

In about 30 days, I will be leaving this world line to return home to 2036. I first want to say thank you for the wonderful conversation and insight into your society. I have learned a great deal and my opinion on quite a few things has changed dramatically.

I will finish the questions that have been posted on this site up to this date. Unfortunately, I must now spend my spare time preparing to leave and I will not be on the computer very much. I do however want to repeat my offer and add a slight twist.

After going over my flight plan home, I have discovered my VGL holdover period is a bit longer than I expected. I will be spending at least three weeks in April of 1998 as I make my way back to 1975. Therefore, I not only offer you the chance to leave a message to yourself in 2036 but I offer you the chance to leave yourself a message in 1998. I will take any compiled messages and email addressees you provide and send them on the net when I get to 1998.

Granted, this will not affect you on your world line now but you make take some comfort that another "you" on another world line has the advantage of knowing something you wish you knew three years ago. Based on the earlier questions I've seen, I've decided a day-to-day record of the

Dow a day in advance should convince you that the messages are real in 1998.

In addition, I am hopeful a series of photocopies and photographs will be available for you that may give you more insight into the technology of the distortion unit. I will let you know the address of the site when it is available. I also plan to have my parents videotape my departure. If they succeed, it will also be posted after I leave.

I look forward to these last few weeks with my family and I will check in periodically to check this site.

Live in Peace 2001,

03-13-2001 04:10 AM

I'm going to try and get to the remaining questions today. (Forum member name) has been collecting the email and forwarding them to another address. In respect for your privacy, I am not reading them. I am only planning to forward them.

03-13-2001 08:41 AM

Bodies under acceleration lose their initial constant velocity world line

reference with respect to each other - the Twins Paradox.

I'm not sure that's accurate. Twin Paradox time travel only suspends your perspective on a local level as the "world" around you goes on. You do not change world lines.

Given that you have a 2.5% divergence from your world line and the micro-singularities are subject to the same divergence, how do you keep them in phase?

Good thinking but that's not exactly the way they work and divergence is not cumulative.

Does the divergence extend into N-dimensions? Is the 2.5% the total error or is each dimension subject to the 2.5% divergence individually?

Yes, that's a little closer. You should perhaps change the "N" to and "X" to avoid string theory confusion.

The "machine" with the energy to do it will come on-line very soon. The "method" for doing it has already been "mostly" perfected in the Z machine at the National lab in New

Mexico.

I believe that it would theoretically take the total energy output of the Sun to form one micro-singularity, let along two.

Not that much.

I haven't seen an answer to my issue concerning moral turpitude through action or inaction. Did I miss it?

If I missed something, please repeat it.

But let me ask you one simple question: instead of sitting at your computer, why not present yourself to the government?

Please take a look at the front cover of this month's (technology magazine) because it's a great example of your legacy to 2036 after the war. One side of the cover it describes in great detail how your government is ready spying on you. On the other side (and just as important) it tells you how to install a hot tub.

The reason time travelers do not revel themselves is because your society scares the hell out of us. We do not want to end up in a cement room on a permanent supply of sodium prenatal as men with lab coats poke at our machine with a screwdriver.

Would not the computer from 1975 be bigger than the time machine used to haul it back to 2036?

Not at all. The 5100 series will fit on a tabletop.

Are people using "reverse speech"?

Not that I'm aware of.

WACO with criminal violations of the law either directly or impliedly as is done in this video, simply doesn't accord with the real facts.

A large point of contention seems to be the "flashes" of light that appear to be gunfire that were recorded from the aircraft flying over the compound. The FBI has stated that these flashes were sunlight reflections. I find that rather interesting since the camera was not a visible light camera, it was a thermal camera. If the federal forces learned anything from WACO it was to install more reliable suppressors on their automatic weapons and don't use flash grenades that leave shell casings after the fire.

Where did you attend High school and what year did you graduate?

No, I did not have a "high school" experience.

What college did you attend, what year did you graduate?

I was educated at the University of Florida. I entered a military sponsored program in 2029 and graduated between 2033 - 34. No, it was not very similar.

If you fell in love with someone here and took her with you in your time machine, wouldn't that act upset both of our world lines?

No, it would not disrupt anyone's world lines.

Have you had a chance to watch a movie here that you had already seen in your 2036 world line? Did they have the same endings?

Interesting question. If I watch enough of them, I should see a difference somewhere but I haven't seen one yet.

03-13-2001 08:46 AM

Where is the mass coming from to form your singularities?

E=MC squared can be solved for mass too.

This appeared to be the same question from the other site so I just copied my old response.

You stated that your society is not involved in space travel. He's also stated that the temperature in and around his device while in use is approximately 100 degrees.

I'm not sure I understand the connection between no space travel and the temperature around the device.

The singularities are not unstable; therefore, uncontrolled evaporation is not possible. In addition, there is no extemporaneous matter near the singularity that would cause it to give off radiation or heat.

Maybe (in order to make a singularity) you've taken a slice of the Earth about 1.2 miles wide and compressed it into a singularity.

A singularity about the size of an electron would only require the mass of a large mountain. The singularities inside the C204 are much small than that. And no, I didn't make them.

Then must be gobbling up the Earth to make their singularities.

You know... E = MC squared can be written to solve for mass too.

03-14-2001 02:05 AM

John Titor, would you consider video taping your departure? What would we expect to be see from the outside perspective?

Earlier in the thread I had said I would be willing to videotape my departure. There are a few technical and logistical problems but I do plan to have it done. (i.e. the videotape recording has static and interference if it's too close to the unit.) At this point, the videotape would be for pure entertainment value. It won't prove one way or another if I'm a time traveler but I feel you deserve just a tad of bread and circuses.

When I approached my grandfather in 1975 it took me quite a while to convince him I was who I said I was. He said something I've never forgotten and I've heard some of you allude to it also. After looking at the unit he turned to me and said, . As the weeks went on, it occurred to me that both were just as threatening and dangerous to him and I'm not sure he ever decided which one was worse.

Based on my own experiences on the web and a few comments some of you have made, I suspect Art is growing weary of people claiming to be time travelers for the same reason. As we have discussed, there is really no way to prove it and I would imagine Art is tired of putting himself at risk by entertaining the idea. He does have a responsibility to his listeners and I respect that. I suppose it goes back to the old question you've all asked yourselves. What is proof of time travel?

03-14-2001 09:23 AM

Why is it not ok to give us information about our near future but it is ok to take back emails and modify another world line?

I'm not saying anything in your messages. You are. Are you suggesting I edit your emails? Are you unable to weigh the consequences of your opportunity and I am now responsible for what you might say to yourself? Now that you have the chance to put your own morals to the test do you feel you're incapable of living up to your own standards?

Is it wrong to say one thing and not something else? If you feel you should tell yourself to buy a certain stock than I suppose you are willing to take the risk that "your" advice doesn't prove wrong in the next few days.

What ever I might do, I would consider the fact that someday you will have to address this question again as an entire society.

Why would I offer to make the video if I thought it would "expose" me? If it makes you feel better, I doubt it will change your mind anyway but it will give you something to talk about when I'm gone. I think that's the greatest gift I could give you.

03-14-2001 11:20 AM

I know you don't understand me, which is sad.

Perhaps you are just having a hard time making yourself clear? I will admit you are a little out of my ballpark but I do understand what you are referring to.

It's Hawking Radiation you can't overcome.

Yes, that is true. If you firmly believe that Hawking radiation cannot be controlled or goes on even without the presence of virtual particles forever until the singularity explodes than you are correct.

A simple "E=Mc^2" isn't the answer.

You asked where the mass comes from. I simply pointed out that mass and energy are interchangeable in the same equation. One of my Stanford pals tells me there is a running gag about the chances a VW Beetle spontaneously appearing inside the accelerator. It could only come from the transfer of energy to mass.

You have to form the singularity for your machine to work and that takes mass.

That is incorrect.

The faulty part of your description for your device involves Hawking Radiation. Its not the size of the singularity that matters, its the mass involved that determines the temperature of the radiation.

You seem to be quite upset and I understand your argument. I do however think it is important to gather the facts and probabilities before expelling emotional energy on them. Please keep in mind that I have not shared all the technical details of the machine with you. So an easy out would be for me to just make something up.

However, and as I'm sure you are aware, Stephen Hawking admits that his own equations support the "possibility" that micro singularities may not totally disappear as they evaporate in a sea of virtual particles and in fact may leave behind a very stable naked singularity. I'm sure you can look that up. I suppose the difficult part is believing that we've taken advantage of it, not that it's impossible.

03-15-2001 01:17 PM

<u>I realize that you haven't given "all of the technical details" of the device.</u>

Actually, I'm hoping the cut-a-way drawing from the manual will be available to you very soon.

<u>If it (the device) exists, the details aren't yours to give in any case.</u>

So let me get this straight, John please prove you're a time traveler but don't show us any copyrighted material?

<u>The details that you have posted publicly may actually be in violation of copyright and patent law.</u>

No, I am not breaking any of "my" laws but I suppose that's something else you and your world line will have to deal with when time travel comes.

<u>The reason that I ask this question is that we don't know that GE isn't working on this device as we speak.</u>

They might be now.

You see a time machine. I see a weapons system.

Yes I suppose that is one thing you could do with it. I could also cut my hand off with a power saw or heat up a crowd of people with a microwave. However, I believe Teller already came up with an X-ray laser that destroys itself after going off.

X-rays will be emitted if matter is pumped into the device.

Actually, I thought we were focusing on the degree of radiation and temperature. I don't believe I ever said it didn't give off radiation. Yes, the device does give off x-rays.

The drawing indicates in Detail #5 "X-Ray Venting Zone". It details x-rays being focused and vented directionally. It has applications as a weapons system.

As you said, it's interesting that I see a time machine and you see a weapon. Maybe it's a sign of the "times". However, it is a good point. If the Chinese or Russians thought you had one of these what do you think they would do?

Again, maybe you should ask yourself if you're sure you want me to prove I'm a time traveler. Maybe that's what makes a time traveler "evil" in that he would be willing to share everything with you.

<u>If John's device is real, it belongs to another world altogether. another GE, therefore it would violate no known copy right laws.</u>

Any government document cannot be copyrighted. I could also argue that the manual "could" be from a future where it has become public domain but then again, it would mean proving I am a time traveler.

03-16-2001 07:09 AM

To my knowledge, there are no other sites where these pictures can be seen and is stable. A few of them have not been posted before. I suspect they will generate more questions which I will try to address.

03-16-2001 03:14 PM

Government documents actually are copyrighted. Here's one example.

My fault. It's Federal documents.

03-17-2001 03:55 AM

Heavier than air aircraft flight happened in 1903. You said 1910. It's only the most important date in the history of aviation and flight other than 1969.

I suppose it's impossible to defend every possible combination of what people want to see. I don't believe I said anything about the date for the first flight. All I did was pick a moment in history.

03-17-2001 05:28 PM

Where mass is accelerated to light speed and forms a singularity doesn't exist.

I can't find where I said that. Could you point that for me?

The speed of light squared is a constant number used to represent the variation between energy and mass. It does not imply that acceleration is required to change or represent the other.

03-18-2001 01:01 AM

<u>The acceleration to light speed is implied in your reference to virtual mass. Virtual particles travel at light speed. I tried to give you an out there but you insisted that the mass was virtual.</u>

The word implied is not a very stable platform to come up with a profile for my parents education but I applaud your attempt.

Well at least we aren't seeing any more thermal and mass stabs in the dark. Interesting profile but you couldn't slide me just 10 more points on the I.Q.?

Are you suggesting that in all cases there must be an acceleration component in the conversion of energy to mass or mass to energy?

03-18-2001 04:27 PM

<u>John's use of the English is very baby-boomer.</u>

I actually worked quite hard on that. It appears the physics questions have come to a halt but at least you're not insulting about my mother anymore. Thanks.

There is nothing in his use of the English language that is atypical of someone born between 1945-1975.

Perhaps you could raise your confidence level to 100% by going from 30 to say... 100 years, maybe 1930 - 2030?

The tools you use to have that much faith in my profile must be pretty good. I'm interested in what you compared me with. How exactly does a person born in 1998 who traveled across world lines from 2036 use the English language?

17 March 2001 11:26

The pictures I promised:

03-24-2001 03:53 PM

I will be leaving this world line shortly and this will be my final post. There are only a handful of people who will know exactly when I will be leaving and I'm sure they will let you know when I'm gone.

In the last few days I have found your choice of topics quite interesting and from an objective viewpoint I think it collectively answers one of your own questions, "If time travel is real, where are all the time travelers?" In the past, I have stated that quite frankly, you all scare the Hell out of me and I'm sure other temporal drivers would feel the same. But now I have an expanded explanation with two examples.

A while ago (on one of the posts), I related an experience I had with my parents while we were driving down a highway. Every now and then, we would pass someone who was in obvious distress with their vehicle. I was amazed that so many people could pass them by without stopping to help. Their explanation was fear. The risk of helping someone was too great and with today's technology, they probably had a cell phone anyway. If they didn't, the walk to a gas station would be good for them and teach them a lesson for running out of gas.

The other example is the plight of the homeless. When you pass them as individuals on the street I see the way people selectively choose an alternate path to avoid them.

Those two examples best define why time travelers do not show themselves. In trying to help you, we put ourselves as great risk and there's really no point to it. We know the nature of time dictates that traveling between "exact" world lines is impossible. Therefore, the only results we will see will be the ones we stay to see. Since world lines, outcomes, and events are infinite, we have better things to do. When I arrive in the "new" 1998 world line on my way home I could easily start all of this again and continue to go through the same conversations with all of the same people. However, I already know you won't pay any attention or believe me because we've already been through it on this world line. Besides, I think the walk to the gas station will do you some good.

The following are the last questions I saw before my "going home" post. I apologize for not being able to get to all of them.

Do most of the people of that time die, especially ones that currently have health problems?

Yes, and people are still dying and a great

deal of them are passing from CJD. As I said, with my very first few posts almost 6 months ago, I want to emphasize how devastating this will be. I believe two people are confirmed dead in Colorado from CJD from surgical instruments. Ahhhh, the power of cheese. Milk does a body good and beef is what's for dinner!

Me: "No, I have not tried any fast food. Thinking about where the food came from, how it was shipped and treated absolutely terrifies me. I have tried to tell people about CJD disease and it seems to be "catching on" in Europe."

Me: Do not eat or use products from any animal that is fed and eats parts of its own dead.

Me: The "Mad Cow" story here is yet to begin...

Is it possible that sometime in your future that time travel will be commonplace?

Yes, that is absolutely possible and eventual.

Have any of the scientists of your time

discovered any new planets, possibly ones with life?

Not that I'm aware of.

Has the bandwidth of the Internet increased greatly?

Yes.

One last question, how did Texas fare during the war?

Texas is still there but Spanish is a lot more popular.

How does time travel affect future exploration of the universe?

There is a great deal of debate about trying to use a distortion unit to "travel" to the moon. The experiment would require very precise calculations that would allow the VGL system to find a theoretical path to the moon on a different world line. The only problem is there is no way to communicate with anyone if the experiment should succeed. In other words, it's possible to do it but only the people on the receiving end could take advantage of it.

My oldest son wanted me to ask if you have any siblings.

No, I am an only child.

Are you still planning on broadcasting your departure over the Internet?

My father will be videotaping the departure. My primary concern is the anonymity of my family. In addition, my departure will be in a somewhat public place and I do not want to draw additional attention to myself.

Is there still (sporting event) in 2036?

Yes, we still play basketball but I am not a fan and can't comment on its organization.

You say that you wear a flight suit and that you experience 2 g's for 6-8 hrs. How is it possible to withstand that kind of g-force for such a long period without the use of an anti-g suit?

The average human can take 2 G's without too much difficulty. Blackout occurs at about 8 or 9 Gs and red-out occurs at negative 3 Gs.

I know you must be physically trained for space travel, but you should also have the benefit of equipment to help you out.

We are encouraged to stay physically fit.

In an effort to address (forum member name) comments, (and in all fairness) the following posts are out of context and not in order. If I'm not a real time traveler I would suggest that Emmett is at a disadvantage on his tactics to "expose" me.

If I present a picture of a sea monster to you and I claim it's real you are forced to argue its validity on the basis of the evidence that I present or create. Under these circumstances, you can't win. If you look at the picture and argue that sea monsters should have more teeth or their incisors are not in proportion to the amount of fish they eat, it's easy to ask you how you know so much about sea monsters if they don't exist. Granted, you could point to dinosaur skeletons and make comparisons. But I can still say, its not a dinosaur, it's a sea monster. In fact, I could even "whip" up some tooth marks on a piece of petrified wood and prove to you don't know anything about sea monsters.

I realize that you haven't given "all of the technical details" of the device but your device, as described, simply won't work.

If you don't have all the details, how do you know it doesn't work?

Honestly, I'm not upset about any of this and the only emotion involved for me is joy. This is fun! It really is.

Joy?

(Forum member name) drew the following assumptions from our conversation and I don't believe I said any of them.

The universe that you described, that is, one where mass is accelerated to light speed and forms a singularity doesn't exist.

As you accelerated to light speed in your machine you and your machine formed a black hole.

From your perspective as you accelerated to light speed every other object in the universe formed a black hole due to your relative velocities.

Take a close look at the technical drawing, very poor quality CAD for a billion dollar project.

A billion? How do you know it's a CAD drawing?

If you have contact with Boomer please attempt to verify the elapsed time for his trip from 2036 to 1975. I need that data for a calculation.

The first leg takes me to 1998. I think I said that quite a few times.

He also said that the mass of the singularities is that of a "small mountain".

If that were true, the unit could not be moved. I only referenced the mass of a small mountain in one of our physics conversations. In fact, I believe I said the mass for the singularities in the distortion unit was much smaller.

The last time I checked I didn't see an "on-off" switch for a black hole.

Me neither. I don't believe I said you could turn it on or off.

At 2 g acceleration you will be traveling just a tad faster than a Chevy pick-up in short order.

This is the one that really disappoints me. Even you should know that Einstein's thought experiment in the isolated elevator was based on the idea that the effects of acceleration and gravity are the same. I never said acceleration had anything to do with how the unit operates.

Does it really make sense that his handlers would authorize a personal vacation with their billion-dollar machine?

I'm looking at my orders and I don't see the word vacation on it anywhere. You had my hopes up for a moment. Interesting how quickly a zillion-dollars has become a fact now.

The answer to the question, "How does a person born in 1998 use the language...?" is - who knows?

I agree. How would you know?

The two yellow caution tapes on the device are misaligned. Sloppy workmanship.

I'll have to point that out to my "handlers" when I get back.

I suppose that if you were interested in verification of Boomer's science that you could go right to the source, Dr. Frank Tipler.

He's a very pleasant gentleman. I highly recommend his book, The Physics of Immortality. I believe I made reference to this earlier.

What would prevent us, the enemy state, from returning the machine to 2036 with a suitcase nuclear device?

This is the second time I've seen a reference to the unit as a weapon. I would submit that the people of this world line have nothing to fear from me. What would you do with the unit?

In your opinion, what is the smallest mass that can form a singularity? Even Hawking suggested primordial singularities were created at the Big Bang. Were there planets (or half planets) around to form them then?

Based on your ability to draw conclusions about someone's profile from their typed words I find it hard to believe you would make such errors in the syntax and meaning of the exact same words. Unfortunately, you have now maneuvered yourself into a position where I have the last word and our debate has come to an end. Boomer isn't a bad name and I sort of like it. Thanks.

If there were an infinite number of realities, then there would necessarily arise a reality that somehow causes there to be no other realities.

Yes!! Excellent insight. I would have enjoyed a conversation on this.

I also want to thank (forum member name) for helping me with the email and everyone else who asked intelligent and insightful questions. I have learned a great deal.

No, I do not have a secret agenda but I have been paying a great deal of attention to your world line. My interaction with you was not a direct mission parameter but it was a secondary mission protocol based on standing orders given to all temporal drivers.

That secondary objective is basically to gather as much information about a world line based on a set of observable variables when we first arrive. Your world line met those conditions. What amazes me is why no one here wonders why Y2K didn't hit them at all?

Bring a gas can with you when the car dies on the side of the road.

Farewell. John

Is this the *REAL* John Titor?

John's family has reportedly moved from Florida – please leave them alone!

Male, 29 years old

Tampa, FL

Jacksonville, FL

Usernames

jtitorisalive • timetravel_0

Work

- Dimension Jumper

Interests

- physics
- hockey
- time travel
- dr. who
- quantum leap
- albert einstein

Social

Twitter @timetravel

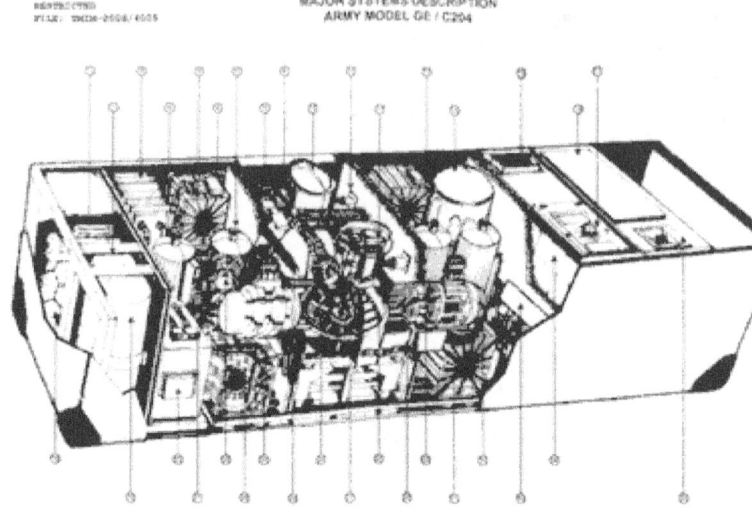

Titor's *Time Machine* Cutaway

I believe this was one of the last pictures John posted and it shows a cut-a-way of his time machine. I tried to look through the posts to find any technical details he mentioned.

The **IBM 5100 Portable Computer** was a portable computer introduced in September 1975, six years before the IBM PC. It was the evolution of a prototype called the **SCAMP** (Special Computer APL Machine Portable) that was developed at the IBM Palo Alto Scientific Center in 1973. In January 1978 IBM announced the IBM 511), its larger cousin, and in February 1980 IBM announced the IBM 5120. The 5100 was withdrawn in March 1982.

When the IBM PC was introduced in 1981, it was originally designated as the IBM 5150, putting it in the "5100" series, though its architecture was not directly descended from the IBM 5100.

Description purportedly from John.
This is a picture taken in the fall of 2035 during my training. It shows my instructor beaming a handheld laser outside the vehicle during operation. The beam is being bent by the gravitational field produced outside the vehicle by the distortion unit. The beam is visible through smoke that is coming from his cigar.

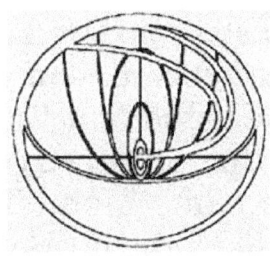

John Titor's *supposed* military insignia.

Please note: John's insignia bears an uncanny resemblance to the German Antarctic Expedition insignia associated with Admiral Byrd 1938-1939. Since I don't believe in coincidences, whether or not there is any connotation it does give me pause – that John Titor is a fake. Like Fox Mulder of the X-Files – *I want to believe*, but I'm no fool, either. *My decision is still out.*

John's first posts appeared on the *Time Travel Institute* forums on November 2, 2000, under the name *TimeTravel_0*. At the time the posts had nothing to do with future events and the name "John Titor" was not being used. Instead, the posts discussed time travel in general, the first one being the "six parts" description of what a time machine would need to have to work (see below) and responses to questions about how such a machine would work. Early messages tended to be short.

The name "John Titor" was not introduced until January 2001, when TimeTravel_0 began posting at the Art Bell BBS Forums (which required a name or pseudonym for every account). The Titor posts ended in late March 2001. Eventually, a number of the threads became corrupted; but Titor's posts had been saved on subscribers' hard drives and were copied to Anomalies.net, along with new discussions of the science behind Titor's time traveling as well as his predictions.

Around 2003, various websites reproduced Titor's posts, re-arranging them into narratives. Not all refer to the original dates posted.

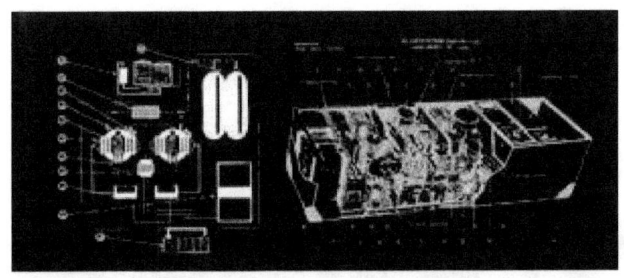

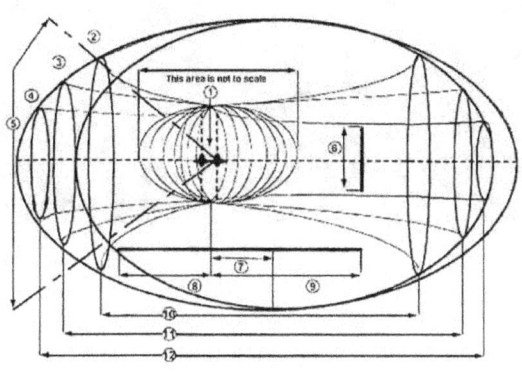

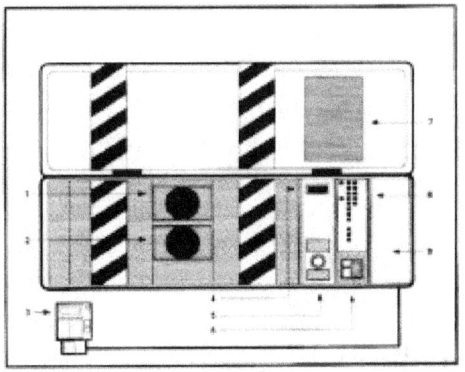

1. MAIN INSTRUMENTS
2. STANDING INSTRUMENTS
3. SORTIE PAPERS UNIT
4. EMERGENCY FLOW CHECK
5. EMERGENCY EQUIPMENT

6. MAIN UNIT INLET SWITCH
7. TEST CONTROL INSTALLATION
8. EMERGENCY CONTROL
9. RADIO STORAGE UNIT

483

◇◇◇

Originally Posted by **Qflux**
John Titor (fax #1) wrote:

Dear Art,

I had to fax when I heard other time travelers calling in from any time past the year 2500 AD. Please let me explain.

Time travel was invented in 2034 off shoots of certain successful fusion reactor research allowed scientists at CERN to produce the worlds first contained singularity engine.

The basic design involves rotating singularities inside a magnetic field. By altering the speed and direction of rotation, you can travel both forward and backward in time.

Time itself can be understood in terms of connected lines. When you go back in time, you travel on your original time line when you turn the singularity engine off a new time line is created due to the fact that you and your time machine are now there.

In other words, a new universe is created. To get back to your original line you must travel a split second farther back and immediately throw the engine into forward without turning it off.

Some interesting outcomes of this are: You meet yourself. I have done it often. Even taken a younger version of myself along for a few rides before returning myself to the new timeline and going back to mine. You can alter history in the new universe that you have just created. Most of the time the changes are subtle. The oldest one was a skyscraper that don't exist in New York. Interestingly when you travel in time, you must compensate for the orbit of the Earth since the time machine doesn't move you have to adjust the engine so you remain on the planet when you turn it off.

Now for the future you might want to know about. Y2K is a disaster. Many people die on the highways when they freeze to death trying to get to warmer weather. The government tries to keep power by instituting marshal law but all of it collapses when their efforts to bring the power back up fail. A few years later communal government system is developed after the constitution takes a few twists. China retakes Tiawan. Isreal wins the largest battle for their life and Russia is covered in

Nuclear snow from their collapsed reactors.

◇◇◇

The Titor Faxes to Art Bell

Coast to Coast AM

Titor first popped up on July 29 1998, through two faxes sent to Art Bell, , host of the overnight talk show Coast to Coast AM. The faxes tell of the discovery of time travel in 2034 and devastation following the Y2K disaster. Art Bell reads the second of the two faxes on air:

Originally Posted by **Qflux**

John Titor (fax #2) wrote:

("Alright I wanted to read you something as I received it. It's from a time traveler." - Art says).

Dear Mr. Bell,

I am glad you 're back. I faxed this information to you the day before you left the air. I wanted to make sure it wasn't lost in the shuffle so I am sending a gift. If ou've already seen this please accept my apologies. If you choose to make this public please do not publish the fax number. I had to fax when I heard the other time traveler calling in from the recent time past, in fact the year 2500 Ad.

Let me explain, Mr. Bell. I sent a fax with this opening on July 29 1998. As I said then I am a time traveler. I have been on this world line since April of this year and I plan to leave soon. Typically, time travelers do not purposely affect the world lines they visit. However, this mission is unusually long and I've grown attached to some of the people I have met here.

("Isn't that the Stockholm syndrome or something?" - Art says).

Anyway. For my own reasons I have decided to help this world line by sharing information about the future with a few people in the hope that it will help their future. I am contacting you for the same reason. Unfortunately, there is no historical reference to your program in my world line.

I beleive you can change your future by creating one now.

Some of the information presented on your program maybe invaluable to up line researchers. I suggest you isolate the programs that concentrate on military technology and new physics theories. Transcribe these programs and put them some place safe away from the box. I recommend some place in the mid west.

("Away from the box? What does he mean?" - Art says).

I also urge you to reconsider your paranoia to the Russians.

("I am not paranoid about the Russians." - Art says).

They are not preparing for war with the average US citizen. They are preparing for war with the US government. They will eventually save this country and the lives of million of Americans.
I realize my claims are a bit difficult to accept so I will send the following once I know you have received this fax. A few pages from the operations manual of my time machine. And a few colored photographs of my vehicle.

("Alright so you know I got it. Send them along." - Art says).

If you wish to contact me, I will be happy to share with you the nature of time, the physics of time travel, and some of the events of your future.

("God I hadn't read this earlier." - Art says).

Please send a return package to...

("And he gives the number." - Art says).

Note: except for the fact a couple named Titor (supposedly John's parents) hired an attorney – we have no certainty the John Titor was the man proclaiming to be the Time Traveler – because when asked if John Titor was his name, he replied - "Yes, John Titor is a real name." This author finds that to be an odd response.

Watchers and Immortals

Other's who *"could be"* Time Traeler

Time Travelers:

A Person who engages in Time Travel.

Watchers:

Watchers are said to be the Fallen Angels (and the "sons of god") also known as the *Beney 'Elohim'*.

Watchers are believed to be humanoids, having attributes both physical and spiritual in nature. They are **specific angelic beings** given charge to watch over humanity: some are good . . . *some are not!*

It's said that Watchers originally were physical beings who attained ascension and immortality upon their home world (some time in eons past) and were given by God the ability to convert themselves to energy for interstellar travel.

Immortals:

Immortals are thought to be individuals who (for some reason) failed a *great task* and were doomed to walk the Earth until the End of Days. Unlike Watchers, they are not angelic, and they typically do not assist mankind in any way.

1) 1941 Bridge Opening

The above photo was allegedly found on the Virtual Museum of Canada website, an online repository of "Canada's rich history and culture." It's said to be of the reopening of the South Fork Bridge in the early 1940s in Gold Bridge, B.C., Canada. But in the photo, something doesn't belong. Who is this strange individual, seemingly out of place - modern attire and all - in what would otherwise be a perfectly ordinary gathering?

2) VonHelton

Vonhelton 1857   Vonhelton 1916   Vonhelton 1945   Present day

Known simply as VonHelton, this enigmatic figure suggests he might be part vampire, but asserts he is 100% time traveler. He has provided proof of his time travel prowess by lining up side-by-side photographs of himself starting in a studio in 1857 England, with stops in 1916 France, 1945 Berlin, and ending in the modern day in front of an American flag. Is he, in fact, a time traveler? Or is his "vampire gene" keeping him immortal?

3) Hakan Nordqvist

Hakan Nordqvist was having a normal if irksome time fixing his leaky sink when he suddenly found himself crawling through a tunnel. At the end of the tunnel was Hakan himself - but as an older man, somewhere around 70. So that his claims wouldn't be refuted, Hakvan filmed himself jovially embracing...himself. The footage even shows the two men showing off their matching tattoos.

4) Time Traveling Charlie Chaplin

Movie goer . . .

A woman filmed outside the 1928 premiere of Charlie Chaplin's The Circus has convinced many people that time travel is real. The unidentified woman holds her hand up to her ear and talks to somebody, although no one is near her. But wait, there were no cell phones in 1928! Could this woman be a time traveler communicating with her contemporaries via mobile phone?

5) Several Faces You Will Recognize.

Imortais, vampiros ou viajantes do tempo?

6) Rapper Jay-Z.

7) Matt Smith

Matt Smith

8) Wanted!

9) Keanu Reeves – (said to openly admit it!)

Parmigianino 1530

Tchaikovsky 1860

Mounet 1875

Reeves 1995

Reeves 2008

Reeves 2011

Keanu Reeves is immortal. He has lived multiple lives, as a composer, actor and artist. He has grown old and died under mysterious circumstances & started over. MINDFUCK

The best way to conceal the truth – is to flaunt it – because no one will believe you!

10. Orlanda Bloom

Nicolae Grigorescu   Orlando Bloom

11) Now . . . look into a mirror. It reflects your image, which is meant to say . . . *the rest of us!*

Our souls are immortal, and this authpor believes in reincarnation. And if you've EVER had a moment in time that seemed unusual: like a hiccup, a repeating, deja vu . . . then in all probability – you have traveled in time: a few seconds, a few minutes, a few days or years – or you now find yourself in an entirely different reality than the one you were born into.

Impossible you say?!

Look around. *Pay attention.* Anybody "dead or alive" *now* – that you remember being "dead or alive" a few days, weeks, months or even years ago?

Exactly!

"Assume nothing! Question everything! Only in these two actions can you be certain of anything!" *Lyn Murray*

Classic Chills
(Supporting the Classics)

The Thing in the Moonlight

H.P. Lovecraft

(1927)

Morgan is not a literary man; in fact he cannot speak English with any degree of coherency. That is what makes me wonder about the words he wrote, though others have laughed. He was alone the evening it happened. Suddenly an unconquerable urge to write came over him, and taking pen in hand he wrote the following:

My name is Howard Phillips. I live at 66 College Street, in Providence, Rhode Island. On November 24, 1927

—

For I know not even what the year may be now, I fell asleep and dreamed, since when I have been unable to awaken. My dream began in a dank, reed choked marsh that lay under a gray autumn sky, with a rugged cliff of lichen crusted stone rising to the north. Impelled by some obscure quest, I ascended a rift or cleft in this beetling precipice, noting as I did so the black mouths of many fearsome burrows extending from both walls into the depths of the stony plateau.

At several points the passage was roofed over by the choking of the upper parts of the narrow fissure; these places being exceeding dark, and forbidding the perception of such burrows as may have existed there. In one such dark space I felt conscious of a singular accession of fright, as if some subtle and bodiless emanation from the abyss were engulfing my spirit; but the blackness was too great for me to perceive the source of my alarm. At length I emerged upon a tableland of moss grown rock and scanty soil, lit by a faint moonlight which had replaced the expiring orb of day. Casting my eyes about, I beheld no living object; but was sensible of a very peculiar stirring far below me, amongst the whispering rushes of the pestilential swamp I had lately quitted.

After walking for some distance, I encountered the rusty tracks of a street railway, and the worm eaten poles which still held the limp and sagging trolley wire. Following this line, I soon came upon a yellow, vestibuled car numbered 1852 of a plain, double trucked type common from 1900 to 1910. It was untenanted, but evidently ready to start; the trolley being on the wire and the air brake now and then throbbing beneath the floor. I boarded it and looked vainly about for the light switch noting as I did so the absence of the controller handle, which thus implied the brief absence of the motorman. Then I sat down in one of the cross seats of the vehicle. Presently I heard a swishing in the sparse grass toward the left, and saw the dark forms of two men looming up in the moonlight. They had the regulation caps of a railway company, and I could not doubt but that they were conductor and motorman. Then one of them sniffedwith singular sharpness, and raised his face to howl to the moon. The other dropped on all fours to run toward the car.

I leaped up at once and raced madly out of that car and across endless l eagues of plateau till exhaustion forced me to stopdoing this not because the conductor had dropped on all fours, but because the face of the motorman was a mere white cone tapering to one blood red tentacle. . . .

I was aware that I only dreamed, but the very awareness was not pleasant. Since that fearful night, I have prayed only for awakening it has not come!Instead I have found myself an inhabitant of this terrible dream world! That first night gave way to dawn, and I wandered aimlessly over the lonely swamp lands. When night came, I still wandered, hoping for awakening. But suddenly I parted the weeds and saw before me the ancient railway car and to one side a cone faced thing lifted its head and in the streaming moonlight howled strangely!

It has been the same each day. Night takes me always to that place of horror. I have tried not moving, with the coming of nightfall, but I must walk in my slumber, for always I awaken with the thing of dread howling before me in the pale moonlight, and I turn and flee madly. God! when will I awaken?

That is what Morgan wrote. I would go to 66 College Street in Providence, but I fear for what I might find there.

The Thing From Space
H.P. Lovecraft
1927

West of Arkham the hills rise wild, and there are valleys with deep woods that no axe has ever cut. There are dark narrow glens where the trees slope fantastically, and where thin brooklets trickle without ever having caught the glint of sunlight. On the gentler slopes there are farms, ancient and rocky, with squat, moss-coated cottages brooding eternally over old New England secrets in the lee of great ledges; but these are all vacant now, the wide chimneys crumbling and the shingled sides bulging perilously beneath low gambrel roofs.

The old folk have gone away, and foreigners do not like to live there. French-Canadians have tried it, Italians have tried it, and the Poles have come and departed. It is not because of anything that can be seen or heard or handled, but because of something that is imagined. The place is not good for the imagination, and does not bring restful dreams at night. It must be this which keeps the foreigners away, for old Ammi Pierce has never told them of anything he recalls from the strange days. Ammi, whose head has been a little queer for years, is the only one who still remains, or who ever talks of the strange days; and he dares to do this because his house is so near the open fields and the travelled roads around Arkham.

There was once a road over the hills and through the valleys, that ran straight where the blasted heath is now; but people ceased to use it and a new road was laid curving far toward the south. Traces of the old one can still be found amidst the weeds of a returning wilderness, and some of them will doubtless linger even when half the hollows are flooded for the new reservoir. Then the dark woods will be cut down and the blasted heath will slumber far below blue waters whose surface will mirror the sky and ripple in the sun. And the secrets of the strange days will be one with the deep's secrets; one with the hidden lore of old ocean, and all the mystery of primal earth.

When I went into the hills and vales to survey for the new reservoir they told me the place was evil. They told me this in Arkham, and because that is a very old town full of witch legends I thought the evil must be something which grandams had whispered to children through centuries. The name "blasted heath" seemed to me very odd and theatrical, and I wondered how it had come into the folklore of a Puritan people. Then I saw that dark westward tangle of glens and slopes for myself, and ceased to wonder at anything besides its own elder mystery. It was morning when I saw it, but shadow lurked always there. The trees grew too thickly, and their trunks were too big for any healthy New England wood. There was too much silence in the dim alleys between them, and the floor was too soft with the dank moss and mattings of infinite years of decay.

In the open spaces, mostly along the line of the old road, there were little hillside farms; sometimes with all the buildings standing, sometimes with only one or two, and sometimes with only a lone chimney or fast-filling cellar. Weeds and briers reigned, and furtive wild things rustled in the undergrowth. Upon everything was a haze of restlessness and oppression; a touch of the unreal and the grotesque, as if some vital element of perspective or chiaroscuro were awry. I did not wonder that the foreigners would not stay, for this was no region to sleep in. It was too much like a landscape of Salvator Rosa; too much like some forbidden woodcut in a tale of terror.

But even all this was not so bad as the blasted heath. I knew it the moment I came upon it at the bottom of a spacious valley; for no other name could fit such a thing, or any other thing fit such a name. It was as if the poet had coined the phrase from having seen this one particular region. It must, I thought as I viewed it, be the outcome of a fire; but why had nothing new ever grown over those five acres of grey desolation that sprawled open to the sky like a great spot eaten by acid in the woods and fields? It lay largely to the north of the ancient road line, but encroached a little on the other side. I felt an odd reluctance about approaching, and did so at last only because my business took me through and past it. There was no vegetation of any kind on that broad expanse, but only a fine grey dust or ash which no wind seemed ever to blow about. The trees near it were sickly and stunted, and many dead trunks stood or lay rotting at the rim. As I walked hurriedly by I saw the tumbled bricks and stones of an old chimney and cellar on my right, and the yawning black maw of an abandoned well whose stagnant vapours played strange tricks with the hues of the sunlight. Even the long, dark woodland climb beyond seemed welcome in contrast, and I marvelled no more at the frightened whispers of Arkham people. There had been no house or ruin near; even in the old days the place must have been lonely and remote. And at twilight, dreading to repass that ominous spot, I walked circuitously back to the town by the curving road on the south. I vaguely wished some clouds would gather, for an odd timidity about the deep skyey voids above had crept into my soul.

In the evening I asked old people in Arkham about the blasted heath, and what was meant by that phrase "strange days" which so many evasively muttered. I could not, however, get any good answers, except that all the mystery was much more recent than I had dreamed. It was not a matter of old legendry at all, but something within the lifetime of those who spoke. It had happened in the 'eighties, and a family had disappeared or was killed. Speakers would not be exact; and because they all told me to pay no attention to old Ammi Pierce's crazy tales, I sought him out the next morning, having heard that he lived alone in the ancient tottering cottage where the trees first begin to get very thick. It was a fearsomely archaic place, and had begun to exude the faint miasmal odour which clings about houses that have stood too long. Only with persistent knocking could I rouse the aged man, and when he shuffled timidly to the door I could tell he was not glad to see me. He was not so feeble as I had expected; but his eyes drooped in a curious way, and his unkempt clothing and white beard made him seem very worn and dismal. Not knowing just how he could best be launched on his tales, I feigned a matter of business; told him of my surveying, and asked vague questions about the district. He was far brighter and more educated than I had been led to think, and before I knew it had grasped quite as much of the subject as any man I had talked with in Arkham. He was not like other rustics I had known in the sections where reservoirs were to be. From him there were no protests at the miles of old wood and farmland to be blotted out, though perhaps there would have been had not his home lain outside the bounds of the future lake. Relief was all that he shewed; relief at the doom of the dark ancient valleys through which he had roamed all his life. They were better under water now—better under water since the strange days. And with this opening his husky voice sank low, while his body leaned forward and his right forefinger began to point shakily and impressively.

It was then that I heard the story, and as the rambling voice scraped and whispered on I shivered again and again despite the summer day. Often I had to recall the speaker from ramblings, piece out scientific points which he knew only by a fading parrot memory of professors' talk, or bridge over gaps where his sense of logic and continuity broke down. When he was done I did not wonder that his mind had snapped a trifle, or that the folk of Arkham would not speak much of the blasted heath. I hurried back before sunset to my hotel, unwilling to have the stars come out above me in the open; and the next day returned to Boston to give up my position. I could not go into that dim chaos of old forest and slope again, or face another time that grey blasted heath where the black well yawned deep beside the tumbled bricks and stones. The reservoir will soon be built now, and all those elder secrets will be safe forever under watery fathoms. But even then I do not believe I would like to visit that country by night—at least, not when the sinister stars are out; and nothing could bribe me to drink the new city water of Arkham.

It all began, old Ammi said, with the meteorite. Before that time there had been no wild legends at all since the witch trials, and even then these western woods were not feared half so much as the small island in the Miskatonic where the devil held court beside a curious stone altar older than the Indians. These were not haunted woods, and their fantastic dusk was never terrible till the strange days. Then there had come that white noontide cloud, that string of explosions in the air, and that pillar of smoke from the valley far in the wood. And by night all Arkham had heard of the great rock that fell out of the sky and bedded itself in the ground beside the well at the Nahum Gardner place. That was the house which had stood where the blasted heath was to come—the trim white Nahum Gardner house amidst its fertile gardens and orchards.

Nahum had come to town to tell people about the stone, and had dropped in at Ammi Pierce's on the way. Ammi was forty then, and

all the queer things were fixed very strongly in his mind. He and his wife had gone with the three professors from Miskatonic University who hastened out the next morning to see the weird visitor from unknown stellar space, and had wondered why Nahum had called it so large the day before. It had shrunk, Nahum said as he pointed out the big brownish mound above the ripped earth and charred grass near the archaic well-sweep in his front yard; but the wise men answered that stones do not shrink. Its heat lingered persistently, and Nahum declared it had glowed faintly in the night. The professors tried it with a geologist's hammer and found it was oddly soft. It was, in truth, so soft as to be almost plastic; and they gouged rather than chipped a specimen to take back to the college for testing. They took it in an old pail borrowed from Nahum's kitchen, for even the small piece refused to grow cool. On the trip back they stopped at Ammi's to rest, and seemed thoughtful when Mrs. Pierce remarked that the fragment was growing smaller and burning the bottom of the pail. Truly, it was not large, but perhaps they had taken less than they thought.

The day after that—all this was in June of '82—the professors had trooped out again in a great excitement. As they passed Ammi's they told him what queer things the specimen had done, and how it had faded wholly away when they put it in a glass beaker. The beaker had gone, too, and the wise men talked of the strange stone's affinity for silicon. It had acted quite unbelievably in that well-ordered laboratory; doing nothing at all and shewing no occluded gases when heated on charcoal, being wholly negative in the borax bead, and soon proving itself absolutely non-volatile at any producible temperature, including that of the oxy-hydrogen blowpipe. On an anvil it appeared highly malleable, and in the dark its luminosity was very marked. Stubbornly refusing to grow cool, it soon had the college in a state of real excitement; and when upon heating before the spectroscope it displayed shining bands unlike any known colours of the normal spectrum there was much breathless talk of new elements, bizarre optical properties, and other things which puzzled men of science are wont to say when faced by the unknown.

Hot as it was, they tested it in a crucible with all the proper reagents. Water did nothing. Hydrochloric acid was the same. Nitric acid and even aqua regia merely hissed and spattered against its torrid invulnerability. Ammi had difficulty in recalling all these things, but recognised some solvents as I mentioned them in the usual order of use. There were ammonia and caustic soda, alcohol and ether, nauseous carbon disulphide and a dozen others; but although the weight grew steadily less as time passed, and the fragment seemed to be slightly cooling, there was no change in the solvents to shew that they had attacked the substance at all. It was a metal, though, beyond a doubt. It was magnetic, for one thing; and after its immersion in the acid solvents there seemed to be faint traces of the Widmannstätten figures found on meteoric iron. When the cooling had grown very considerable, the testing was carried on in glass; and it was in a glass beaker that they left all the chips made of the original fragment during the work. The next morning both chips and beaker were gone without trace, and only a charred spot marked the place on the wooden shelf where they had been.

All this the professors told Ammi as they paused at his door, and once more he went with them to see the stony messenger from the stars, though this time his wife did not accompany him. It had now most certainly shrunk, and even the sober professors could not doubt the truth of what they saw. All around the dwindling brown lump near the well was a vacant space, except where the earth had caved in; and whereas it had been a good seven feet across the day before, it was now scarcely five. It was still hot, and the sages studied its surface curiously as they detached another and larger piece with hammer and chisel. They gouged deeply this time, and as they pried away the smaller mass they saw that the core of the thing was not quite homogeneous.

They had uncovered what seemed to be the side of a large coloured globule imbedded in the substance. The colour, which resembled some of the bands in the meteor's strange spectrum, was almost impossible to describe; and it was only by analogy that they called it colour at all. Its texture was glossy, and upon tapping it appeared to promise both brittleness and hollowness. One of the professors gave it a smart blow with a hammer, and it burst with a nervous little pop. Nothing was emitted, and all trace of the thing vanished with the puncturing. It left behind a hollow spherical space about three inches across, and all thought it probable that others would be discovered as the enclosing substance wasted away.

Conjecture was vain; so after a futile attempt to find additional globules by drilling, the seekers left again with their new specimen—which proved, however, as baffling in the laboratory as its predecessor had been.

Aside from being almost plastic, having heat, magnetism, and slight luminosity, cooling slightly in powerful acids, possessing an unknown spectrum, wasting away in air, and attacking silicon compounds with mutual destruction as a result, it presented no identifying features whatsoever; and at the end of the tests the college scientists were forced to own that they could not place it. It was nothing of this earth, but a piece of the great outside; and as such dowered with outside properties and obedient to outside laws.

That night there was a thunderstorm, and when the professors went out to Nahum's the next day they met with a bitter disappointment. The stone, magnetic as it had been, must have had some peculiar electrical property; for it had "drawn the lightning", as Nahum said, with a singular persistence. Six times within an hour the farmer saw the lightning strike the furrow in the front yard, and when the storm was over nothing remained but a ragged pit by the ancient well-sweep, half-choked with caved-in earth. Digging had borne no fruit, and the scientists verified the fact of the utter vanishment. The failure was total; so that nothing was left to do but go back to the laboratory and test again the disappearing fragment left carefully cased in lead. That fragment lasted a week, at the end of which nothing of value had been learned of it. When it had gone, no residue was left behind, and in time the professors felt scarcely sure they had indeed seen with waking eyes that cryptic vestige of the fathomless gulfs outside; that lone, weird message from other universes and other realms of matter, force, and entity.

As was natural, the Arkham papers made much of the incident with its collegiate sponsoring, and sent reporters to talk with Nahum Gardner and his family. At least one Boston daily also sent a scribe, and Nahum quickly became a kind of local celebrity. He was a lean, genial person of about fifty, living with his wife and three sons on the pleasant farmstead in the valley. He and Ammi exchanged visits frequently, as did their wives; and Ammi had nothing but praise for him after all these years. He seemed slightly proud of the notice his place had attracted, and talked often of the meteorite in the succeeding weeks. That July and August were hot, and Nahum worked hard at his haying in the ten-acre pasture across Chapman's Brook; his rattling wain wearing deep ruts in the shadowy lanes between. The labour tired him more than it had in other years, and he felt that age was beginning to tell on him.

Then fell the time of fruit and harvest. The pears and apples slowly ripened, and Nahum vowed that his orchards were prospering as never before. The fruit was growing to phenomenal size and unwonted gloss, and in such abundance that extra barrels were ordered to handle the future crop. But with the ripening came sore disappointment; for of all that gorgeous array of specious lusciousness not one single jot was fit to eat. Into the fine flavour of the pears and apples had crept a stealthy bitterness and sickishness, so that even the smallest of bites induced a lasting disgust. It was the same with the melons and tomatoes, and Nahum sadly saw that his entire crop was lost. Quick to connect events, he declared that the meteorite had poisoned the soil, and thanked heaven that most of the other crops were in the upland lot along the road.

Winter came early, and was very cold. Ammi saw Nahum less often than usual, and observed that he had begun to look worried. The rest of his family, too, seemed to have grown taciturn; and were far from steady in their churchgoing or their attendance at the various social events of the countryside. For this reserve or melancholy no cause could be found, though all the household confessed now and then to poorer health and a feeling of vague disquiet. Nahum himself gave the most definite statement of anyone when he said he was disturbed about certain footprints in the snow. They were the usual winter prints of red squirrels, white rabbits, and foxes, but the brooding farmer professed to see something not quite right about their nature and arrangement. He was never specific, but appeared to think that they were not as characteristic of the anatomy and habits of squirrels and rabbits and foxes as they ought to be. Ammi listened without interest to this talk until one night when he drove past Nahum's house in his sleigh on the way back from Clark's Corners. There had been a moon, and a rabbit had run across the road, and the leaps of that rabbit were longer than either Ammi or his horse liked. The latter, indeed, had almost run away when brought up by a firm rein. Thereafter Ammi gave Nahum's tales more respect, and wondered why the Gardner dogs seemed so cowed and quivering every morning. They had, it developed, nearly lost the spirit to bark.

In February the McGregor boys from Meadow Hill were out shooting woodchucks, and not far from the Gardner place bagged a very peculiar specimen. The proportions of its body seemed slightly altered in a queer way impossible to describe, while its face had taken on an expression which no one ever saw in a woodchuck before. The boys were genuinely frightened, and threw the thing away at once, so that only their grotesque tales of it ever reached the people of the countryside. But the shying of the horses near Nahum's house had now become an acknowledged thing, and all the basis for a cycle of whispered legend was fast taking form.

People vowed that the snow melted faster around Nahum's than it did anywhere else, and early in March there was an awed discussion in Potter's general store at Clark's Corners. Stephen Rice had driven past Gardner's in the morning, and had noticed the skunk-cabbages coming up through the mud by the woods across the road. Never were things of such size seen before, and they held strange colours that could not be put into any words. Their shapes were monstrous, and the horse had snorted at an odour which struck Stephen as wholly unprecedented. That afternoon several persons drove past to see the abnormal growth, and all agreed that plants of that kind ought never to sprout in a healthy world. The bad fruit of the fall before was freely mentioned, and it went from mouth to mouth that there was poison in Nahum's ground. Of course it was the meteorite; and remembering how strange the men from the college had found that stone to be, several farmers spoke about the matter to them.

One day they paid Nahum a visit; but having no love of wild tales and folklore were very conservative in what they inferred. The plants were certainly odd, but all skunk-cabbages are more or less odd in shape and odour and hue. Perhaps some mineral element from the stone had entered the soil, but it would soon be washed away. And as for the footprints and frightened horses—of course this was mere country talk which such a phenomenon as the aërolite would be certain to start. There was really nothing for serious men to do in cases of wild gossip, for superstitious rustics will say and believe anything. And so all through the strange days the professors stayed away in contempt. Only one of them, when given two phials of dust for analysis in a police job over a year and a half later, recalled that the queer colour of that skunk-cabbage had been very like one of the anomalous bands of light shewn by the meteor fragment in the college spectroscope, and like the brittle globule found imbedded in the stone from the abyss. The samples in this analysis case gave the same odd bands at first, though later they lost the property.

The trees budded prematurely around Nahum's, and at night they swayed ominously in the wind. Nahum's second son Thaddeus, a lad of fifteen, swore that they swayed also when there was no wind; but even the gossips would not credit this. Certainly, however, restlessness was in the air. The entire Gardner family developed the habit of stealthy listening, though not for any sound which they could consciously name. The listening was, indeed, rather a product of moments when consciousness seemed half to slip away. Unfortunately such moments increased week by week, till it became common speech that "something was wrong with all Nahum's folks". When the early saxifrage came out it had another strange colour; not quite like that of the skunk-cabbage, but plainly related and equally unknown to anyone who saw it. Nahum took some blossoms to Arkham and shewed them to the editor of the *Gazette,* but that dignitary did no more than write a humorous article about them, in which the dark fears of rustics were held up to polite ridicule. It was a mistake of Nahum's to tell a stolid city man about the way the great, overgrown mourning-cloak butterflies behaved in connexion with these saxifrages.

April brought a kind of madness to the country folk, and began that disuse of the road past Nahum's which led to its ultimate abandonment. It was the vegetation. All the orchard trees blossomed forth in strange colours, and through the stony soil of the yard and adjacent pasturage there sprang up a bizarre growth which only a botanist could connect with the proper flora of the region. No sane wholesome colours were anywhere to be seen except in the green grass and leafage; but everywhere those hectic and prismatic variants of some diseased, underlying primary tone without a place among the known tints of earth. The Dutchman's breeches became a thing of sinister menace, and the bloodroots grew insolent in their chromatic perversion. Ammi and the Gardners thought that most of the colours had a sort of haunting familiarity, and decided that they reminded one of the brittle globule in the meteor. Nahum ploughed and sowed the ten-acre pasture and the upland lot, but did nothing with the land around the house. He knew it would be of no use, and hoped that the summer's strange growths would draw all the poison from the soil. He was prepared for almost anything now, and had grown used to the sense of something near him waiting to be heard. The shunning of his house by neighbours told on him, of course; but it told on his wife more. The boys were better off, being at school each day; but they could not help being frightened by the gossip. Thaddeus, an especially sensitive youth, suffered the most.

In May the insects came, and Nahum's place became a nightmare of buzzing and crawling. Most of the creatures seemed not quite usual in their aspects and motions, and their nocturnal habits contradicted all former experience. The Gardners took to watching at night—watching in all directions at random for something . . . they could not tell what. It was then that they all owned that Thaddeus had been right about the trees. Mrs. Gardner was the next to see it from the window as she watched the swollen boughs of a maple against a moonlit sky. The boughs surely moved, and there was no wind. It must be the sap. Strangeness had come into everything growing now. Yet it was none of Nahum's family at all who made the next discovery. Familiarity had dulled them, and what they could not see was glimpsed by a timid windmill salesman from Bolton who drove by one night in ignorance of the country legends. What he told in Arkham was given a short paragraph in the *Gazette;* and it was there that all the farmers, Nahum included, saw it first. The night had been dark and the buggy-lamps faint, but around a farm in the valley which everyone knew from the account must be Nahum's the darkness had been less thick.

A dim though distinct luminosity seemed to inhere in all the vegetation, grass, leaves, and blossoms alike, while at one moment a detached piece of the phosphorescence appeared to stir furtively in the yard near the barn.

The grass had so far seemed untouched, and the cows were freely pastured in the lot near the house, but toward the end of May the milk began to be bad. Then Nahum had the cows driven to the uplands, after which the trouble ceased. Not long after this the change in grass and leaves became apparent to the eye. All the verdure was going grey, and was developing a highly singular quality of brittleness. Ammi was now the only person who ever visited the place, and his visits were becoming fewer and fewer. When school closed the Gardners were virtually cut off from the world, and sometimes let Ammi do their errands in town. They were failing curiously both physically and mentally, and no one was surprised when the news of Mrs. Gardner's madness stole around.

It happened in June, about the anniversary of the meteor's fall, and the poor woman screamed about things in the air which she could not describe. In her raving there was not a single specific noun, but only verbs and pronouns. Things moved and changed and fluttered, and ears tingled to impulses which were not wholly sounds. Something was taken away—she was being drained of something— something was fastening itself on her that ought not to be—someone must make it keep off—nothing was ever still in the night—the walls and windows shifted. Nahum did not send her to the county asylum, but let her wander about the house as long as she was harmless to herself and others. Even when her expression changed he did nothing. But when the boys grew afraid of her, and Thaddeus nearly fainted at the way she made faces at him, he decided to keep her locked in the attic. By July she had ceased to speak and crawled on all fours, and before that month was over Nahum got the mad notion that she was slightly luminous in the dark, as he now clearly saw was the case with the nearby vegetation.

It was a little before this that the horses had stampeded. Something had aroused them in the night, and their neighing and kicking in their stalls had been terrible. There seemed virtually nothing to do to calm them, and when Nahum opened the stable door they all bolted out like frightened woodland deer. It took a week to track all four, and when found they were seen to be quite useless and unmanageable. Something had snapped in their brains, and each one had to be shot for its own good. Nahum borrowed a horse from Ammi for his haying, but found it would not approach the barn. It shied, balked, and whinnied, and in the end he could do nothing but drive it into the yard while the men used their own strength to get the heavy wagon near enough the hayloft for convenient pitching. And all the while the vegetation was turning grey and brittle. Even the flowers whose hues had been so strange were greying now, and the fruit was coming out grey and dwarfed and tasteless. The asters and goldenrod bloomed grey and distorted, and the roses and zinneas and hollyhocks in the front yard were such blasphemous-looking things that Nahum's oldest boy Zenas cut them down. The strangely puffed insects died about that time, even the bees that had left their hives and taken to the woods.

By September all the vegetation was fast crumbling to a greyish powder, and Nahum feared that the trees would die before the poison was out of the soil. His wife now had spells of terrific screaming, and he and the boys were in a constant state of nervous tension. They shunned people now, and when school opened the boys did not go. But it was Ammi, on one of his rare visits, who first realised that the well water was no longer good. It had an evil taste that was not exactly foetid nor exactly salty, and Ammi advised his friend to dig another well on higher ground to use till the soil was good again. Nahum, however, ignored the warning, for he had by that time become calloused to strange and unpleasant things. He and the boys continued to use the tainted supply, drinking it as listlessly and mechanically as they ate their meagre and ill-cooked meals and did their thankless and monotonous chores through the aimless days. There was something of stolid resignation about them all, as if they walked half in another world between lines of nameless guards to a certain and familiar doom.

Thaddeus went mad in September after a visit to the well. He had gone with a pail and had come back empty-handed, shrieking and waving his arms, and sometimes lapsing into an inane titter or a whisper about "the moving colours down there". Two in one family was pretty bad, but Nahum was very brave about it. He let the boy run about for a week until he began stumbling and hurting himself, and then he shut him in an attic room across the hall from his mother's. The way they screamed at each other from behind their locked doors was very terrible, especially to little Merwin, who fancied they talked in some terrible language that was not of earth. Merwin was getting frightfully imaginative, and his restlessness was worse after the shutting away of the brother who had been his greatest playmate.

Almost at the same time the mortality among the livestock commenced. Poultry turned greyish and died very quickly, their meat being found dry and noisome upon cutting. Hogs grew inordinately fat, then suddenly began to undergo loathsome changes which no one could explain. Their meat was of course useless, and Nahum was at his wit's end. No rural veterinary would approach his place, and the city veterinary from Arkham was openly baffled. The swine began growing grey and brittle and falling to pieces before they died, and their eyes and muzzles developed singular alterations. It was very inexplicable, for they had never been fed from the tainted vegetation. Then something struck the cows. Certain areas or sometimes the whole body would be uncannily shrivelled or compressed, and atrocious collapses or disintegrations were common. In the last stages—and death was always the result—there would be a greying and turning brittle like that which beset the hogs. There could be no question of poison, for all the cases occurred in a locked and undisturbed barn. No bites of prowling things could have brought the virus, for what live beast of earth can pass through solid obstacles? It must be only natural disease—yet what disease could wreak such results was beyond any mind's guessing. When the harvest came there was not an animal surviving on the place, for the stock and poultry were dead and the dogs had run away. These dogs, three in number, had all vanished one night and were never heard of again. The five cats had left some time before, but their going was scarcely noticed since there now seemed to be no mice, and only Mrs. Gardner had made pets of the graceful felines.

On the nineteenth of October Nahum staggered into Ammi's house with hideous news. The death had come to poor Thaddeus in his attic room, and it had come in a way which could not be told. Nahum had dug a grave in the railed family plot behind the farm, and had put therein what he found. There could have been nothing from outside, for the small barred window and locked door were intact; but it was much as it had been in the barn. Ammi and his wife consoled the stricken man as best they could, but shuddered as they did so. Stark terror seemed to cling round the Gardners and all they touched, and the very presence of one in the house was a breath from regions unnamed and unnamable. Ammi accompanied Nahum home with the greatest reluctance, and did what he might to calm the hysterical sobbing of little Merwin. Zenas needed no calming. He had come of late to do nothing but stare into space and obey what his father told him; and Ammi thought that his fate was very merciful. Now and then Merwin's screams were answered faintly from the attic, and in response to an inquiring look Nahum said that his wife was getting very feeble. When night approached, Ammi managed to get away; for not even friendship could make him stay in that spot when the faint glow of the vegetation began and the trees may or may not have swayed without wind. It was really lucky for Ammi that he was not more imaginative. Even as things were, his mind was bent ever so slightly; but had he been able to connect and reflect upon all the portents around him he must inevitably have turned a total maniac. In the twilight he hastened home, the screams of the mad woman and the nervous child ringing horribly in his ears.

Three days later Nahum lurched into Ammi's kitchen in the early morning, and in the absence of his host stammered out a desperate tale once more, while Mrs. Pierce listened in a clutching fright. It was little Merwin this time. He was gone. He had gone out late at night with a lantern and pail for water, and had never come back. He'd been going to pieces for days, and hardly knew what he was about. Screamed at everything. There had been a frantic shriek from the yard then, but before the father could get to the door, the boy was gone. There was no glow from the lantern he had taken, and of the child himself no trace. At the time Nahum thought the lantern and pail were gone too; but when dawn came, and the man had plodded back from his all-night search of the woods and fields, he had found some very curious things near the well. There was a crushed and apparently somewhat melted mass of iron which had certainly been the lantern; while a bent bail and twisted iron hoops beside it, both half-fused, seemed to hint at the remnants of the pail. That was all. Nahum was past imagining, Mrs. Pierce was blank, and Ammi, when he had reached home and heard the tale, could give no guess. Merwin was gone, and there would be no use in telling the people around, who shunned all Gardners now. No use, either, in telling the city people at Arkham who laughed at everything. Thad was gone, and now Merwin was gone. Something was creeping and creeping and waiting to be seen and felt and heard. Nahum would go soon, and he wanted Ammi to look after his wife and Zenas if they survived him. It must all be a judgment of some sort; though he could not fancy what for, since he had always walked uprightly in the Lord's ways so far as he knew.

For over two weeks Ammi saw nothing of Nahum; and then, worried about what might have happened, he overcame his fears and paid the Gardner place a visit. There was no smoke from the great chimney, and for a moment the visitor was apprehensive of the worst. The aspect of the whole farm was shocking—greyish withered grass and leaves on the ground, vines falling in brittle wreckage from archaic walls and gables, and great bare trees clawing up at the grey November sky with a studied malevolence which Ammi could not but feel had come from some subtle change in the tilt of the branches. But Nahum was alive, after all. He was weak, and lying on a couch in the low-ceiled kitchen, but perfectly conscious and able to give simple orders to Zenas. The room was deadly cold; and as Ammi visibly shivered, the host shouted huskily to Zenas for more wood. Wood, indeed, was sorely needed; since the cavernous fireplace was unlit and empty, with a cloud of soot blowing about in the chill wind that came down the chimney. Presently

Nahum asked him if the extra wood had made him any more comfortable, and then Ammi saw what had happened. The stoutest cord had broken at last, and the hapless farmer's mind was proof against more sorrow.

Questioning tactfully, Ammi could get no clear data at all about the missing Zenas. "In the well—he lives in the well—" was all that the clouded father would say. Then there flashed across the visitor's mind a sudden thought of the mad wife, and he changed his line of inquiry. "Nabby? Why, here she is!" was the surprised response of poor Nahum, and Ammi soon saw that he must search for himself. Leaving the harmless babbler on the couch, he took the keys from their nail beside the door and climbed the creaking stairs to the attic. It was very close and noisome up there, and no sound could be heard from any direction. Of the four doors in sight, only one was locked, and on this he tried various keys on the ring he had taken. The third key proved the right one, and after some fumbling Ammi threw open the low white door.

It was quite dark inside, for the window was small and half-obscured by the crude wooden bars; and Ammi could see nothing at all on the wide-planked floor. The stench was beyond enduring, and before proceeding further he had to retreat to another room and return with his lungs filled with breathable air. When he did enter he saw something dark in the corner, and upon seeing it more clearly he screamed outright. While he screamed he thought a momentary cloud eclipsed the window, and a second later he felt himself brushed as if by some hateful current of vapour. Strange colours danced before his eyes; and had not a present horror numbed him he would have thought of the globule in the meteor that the geologist's hammer had shattered, and of the morbid vegetation that had sprouted in the spring. As it was he thought only of the blasphemous monstrosity which confronted him, and which all too clearly had shared the nameless fate of young Thaddeus and the livestock. But the terrible thing about this horror was that it very slowly and perceptibly moved as it continued to crumble.

Ammi would give me no added particulars to this scene, but the shape in the corner does not reappear in his tale as a moving object. There are things which cannot be mentioned, and what is done in common humanity is sometimes cruelly judged by the law. I gathered that no moving thing was left in that attic room, and that to leave anything capable of motion there would have been a deed so monstrous as to damn any accountable being to eternal torment. Anyone but a stolid farmer would have fainted or gone mad, but Ammi walked conscious through that low doorway and locked the accursed secret behind him. There would be Nahum to deal with now; he must be fed and tended, and removed to some place where he could be cared for.

Commencing his descent of the dark stairs, Ammi heard a thud below him. He even thought a scream had been suddenly choked off, and recalled nervously the clammy vapour which had brushed by him in that frightful room above. What presence had his cry and entry started up? Halted by some vague fear, he heard still further sounds below. Indubitably there was a sort of heavy dragging, and a most detestably sticky noise as of some fiendish and unclean species of suction. With an associative sense goaded to feverish heights, he thought unaccountably of what he had seen upstairs. Good God! What eldritch dream-world was this into which he had blundered? He dared move neither backward nor forward, but stood there trembling at the black curve of the boxed-in staircase. Every trifle of the scene burned itself into his brain. The sounds, the sense of dread expectancy, the darkness, the steepness of the narrow steps—and merciful heaven! . . . the faint but unmistakable luminosity of all the woodwork in sight; steps, sides, exposed laths, and beams alike!

Then there burst forth a frantic whinny from Ammi's horse outside, followed at once by a clatter which told of a frenzied runaway. In another moment horse and buggy had gone beyond earshot, leaving the frightened man on the dark stairs to guess what had sent them. But that was not all. There had been another sound out there. A sort of liquid splash—water—it must have been the well. He had left Hero untied near it, and a buggy-wheel must have brushed the coping and knocked in a stone. And still the pale phosphorescence glowed in that detestably ancient woodwork. God! how old the house was! Most of it built before 1670, and the gambrel roof not later than 1730.

A feeble scratching on the floor downstairs now sounded distinctly, and Ammi's grip tightened on a heavy stick he had picked up in the attic for some purpose. Slowly nerving himself, he finished his descent and walked boldly toward the kitchen. But he did not complete the walk, because what he sought was no longer there. It had come to meet him, and it was still alive after a fashion. Whether it had crawled or whether it had been dragged by any external force, Ammi could not say; but the death had been at it. Everything had happened in the last half-hour, but collapse, greying, and disintegration were already far advanced. There was a horrible brittleness, and dry fragments were scaling off. Ammi could not touch it, but looked horrifiedly into the distorted parody that had been a face. "What was it, Nahum—what was it?" he whispered, and the cleft, bulging lips were just able to crackle out a final answer.

"Nothin' . . . nothin' . . . the colour . . . it burns . . . cold an' wet . . . but it burns . . . it lived in the well . . . I seen it . . . a kind o' smoke . . . jest like the flowers last spring . . . the well shone at night . . . Thad an' Mernie an' Zenas . . . everything alive . . . suckin' the life out of everything . . . in that stone . . . it must a' come in that stone . . . pizened the whole place . . . dun't know what it wants . . . that round thing them men from the college dug outen the stone . . . they smashed it . . . it was that same colour . . . jest the same, like the flowers an' plants . . . must a' ben more of 'em . . . seeds . . . seeds . . . they growed . . . I seen it the fust time this week . . . must a' got strong on Zenas . . . he was a big boy, full o' life . . . it beats down your mind an' then gits ye . . . burns ye up . . . in the well water . . . you was right about that . . . evil water . . . Zenas never come back from the well . . . can't git away . . . draws ye . . . ye know summ'at's comin', but 'tain't no use . . . I seen it time an' agin senct Zenas was took . . . whar's Nabby, Ammi? . . . my head's no good . . . dun't know how long senct I fed her . . . it'll git her ef we ain't keerful . . . jest a colour . . . her face is gettin' to hev that colour sometimes towards night . . . an' it burns an' sucks . . . it come from some place whar things ain't as they is here . . . one o' them professors said so . . . he was right . . . look out, Ammi, it'll do suthin' more . . . sucks the life out. . . ."

But that was all. That which spoke could speak no more because it had completely caved in. Ammi laid a red checked tablecloth over what was left and reeled out the back door into the fields. He climbed the slope to the ten-acre pasture and stumbled home by the north road and the woods. He could not pass that well from which his horse had run away. He had looked at it through the window, and had seen that no stone was missing from the rim. Then the lurching buggy had not dislodged anything after all—the splash had been something else—something which went into the well after it had done with poor Nahum. . . .

When Ammi reached his house the horse

and buggy had arrived before him and thrown his wife into fits of anxiety. Reassuring her without explanations, he set out at once for Arkham and notified the authorities that the Gardner family was no more. He indulged in no details, but merely told of the deaths of Nahum and Nabby, that of Thaddeus being already known, and mentioned that the cause seemed to be the same strange ailment which had killed the livestock. He also stated that Merwin and Zenas had disappeared. There was considerable questioning at the police station, and in the end Ammi was compelled to take three officers to the Gardner farm, together with the coroner, the medical examiner, and the veterinary who had treated the diseased animals. He went much against his will, for the afternoon was advancing and he feared the fall of night over that accursed place, but it was some comfort to have so many people with him.

The six men drove out in a democrat-wagon, following Ammi's buggy, and arrived at the pest-ridden farmhouse about four o'clock. Used as the officers were to gruesome experiences, not one remained unmoved at what was found in the attic and under the red checked tablecloth on the floor below. The whole aspect of the farm with its grey desolation was terrible enough, but those two crumbling objects were beyond all bounds. No one could look long at them, and even the medical examiner admitted that there was very little to examine. Specimens could be analysed, of course, so he busied himself in obtaining them—and here it develops that a very puzzling aftermath occurred at the college laboratory where the two phials of dust were finally taken. Under the spectroscope both samples gave off an unknown spectrum, in which many of the baffling bands were precisely like those which the strange meteor had yielded in the previous year. The property of emitting this spectrum vanished in a month, the dust thereafter consisting mainly of alkaline phosphates and carbonates.

Ammi would not have told the men about the well if he had thought they meant to do anything then and there. It was getting toward sunset, and he was anxious to be away. But he could not help glancing nervously at the stony curb by the great sweep, and when a detective questioned him he admitted that Nahum had feared something down there—so much so that he had never even thought of searching it for Merwin or Zenas. After that nothing would do but that they empty and explore the well immediately, so Ammi had to wait trembling while pail after pail of rank water was hauled up and splashed on the soaking ground outside. The men sniffed in disgust at the fluid, and toward the last held their noses against the foetor they were uncovering. It was not so long a job as they had feared it would be, since the water was phenomenally low. There is no need to speak too exactly of what they found. Merwin and Zenas were both there, in part, though the vestiges were mainly skeletal. There were also a small deer and a large dog in about the same state, and a number of bones of smaller animals. The ooze and slime at the bottom seemed inexplicably porous and bubbling, and a man who descended on hand-holds with a long pole found that he could sink the wooden shaft to any depth in the mud of the floor without meeting any solid obstruction.

Twilight had now fallen, and lanterns were brought from the house. Then, when it was seen that nothing further could be gained from the well, everyone went indoors and conferred in the ancient sitting-room while the intermittent light of a spectral half-moon played wanly on the grey desolation outside. The men were frankly nonplussed by the entire case, and could find no convincing common element to link the strange vegetable conditions, the unknown disease of livestock and humans, and the unaccountable deaths of Merwin and Zenas in the tainted well. They had heard the common country talk, it is true; but could not believe that anything contrary to natural law had occurred. No doubt the meteor had poisoned the soil, but the illness of persons and animals who had eaten nothing grown in that soil was another matter. Was it the well water? Very possibly. It might be a good idea to analyse it. But what peculiar madness could have made both boys jump into the well? Their deeds were so similar— and the fragments shewed that they had both suffered from the grey brittle death. Why was everything so grey and brittle?

It was the coroner, seated near a window overlooking the yard, who first noticed the glow about the well. Night had fully set in, and all the abhorrent grounds seemed faintly luminous with more than the fitful moonbeams; but this new glow was something definite and distinct, and appeared to shoot up from the black pit like a softened ray from a searchlight, giving dull reflections in the little ground pools where the water had been emptied. It had a very queer colour, and as all the men clustered round the window Ammi gave a violent start. For this strange beam of ghastly miasma was to him of no unfamiliar hue. He had seen that colour before, and feared to think what it might mean. He had seen it in the nasty brittle globule in that aërolite two summers ago, had seen it in the crazy vegetation of the springtime, and had thought he had seen it for an instant that very morning against the small barred window of that terrible attic room where nameless things had happened. It had flashed there a second, and a clammy and hateful current of vapour had brushed past him—and then poor Nahum had been taken by something of that colour. He had said so at the last—said it was the globule and the plants. After that had come the runaway in the yard and the splash in the well—and now that well was belching forth to the night a pale insidious beam of the same daemoniac tint.

It does credit to the alertness of Ammi's mind that he puzzled even at that tense moment over a point which was essentially scientific. He could not but wonder at his gleaning of the same impression from a vapour glimpsed in the daytime, against a window opening on the morning sky, and from a nocturnal exhalation seen as a phosphorescent mist against the black and blasted landscape. It wasn't right—it was against Nature—and he thought of those terrible last words of his stricken friend, "It come from some place whar things ain't as they is here . . . one o' them professors said so. . . ."

All three horses outside, tied to a pair of shrivelled saplings by the road, were now neighing and pawing frantically. The wagon driver started for the door to do something, but Ammi laid a shaky hand on his shoulder. "Dun't go out thar," he whispered. "They's more to this nor what we know. Nahum said somethin' lived in the well that sucks your life out. He said it must be some'at growed from a round ball like one we all seen in the meteor stone that fell a year ago June. Sucks an' burns, he said, an' is jest a cloud of colour like that light out thar now, that ye can hardly see an' can't tell what it is. Nahum thought it feeds on everything livin' an' gits stronger all the time. He said he seen it this last week. It must be somethin' from away off in the sky like the men from the college last year says the meteor stone was. The way it's made an' the way it works ain't like no way o' God's world. It's some'at from beyond."

So the men paused indecisively as the light from the well grew stronger and the hitched horses pawed and whinnied in increasing frenzy. It was truly an awful moment; with terror in that ancient and accursed house itself, four monstrous sets of fragments—two from the house and two from the well—in the woodshed behind, and that shaft of unknown and unholy iridescence from the slimy depths in front. Ammi had restrained the driver on impulse, forgetting how uninjured he himself was after the clammy brushing of that coloured vapour in the attic room, but perhaps it is just as well that he acted as he did. No one will ever know what was abroad that night; and though the blasphemy from beyond had not so far hurt any human of unweakened mind, there is no telling what it might not have done at that last moment, and with its seemingly increased strength and the special signs of purpose it was soon to display beneath the half-clouded moonlit sky.

All at once one of the detectives at the window gave a short, sharp gasp. The others looked at him, and then quickly followed his own gaze upward to the point at which its idle straying had been suddenly arrested. There was no need for words. What had been disputed in country gossip was disputable no longer, and it is because of the thing which every man of that party agreed in whispering later on that the strange days are never talked about in Arkham. It is necessary to premise that there was no wind at that hour of the evening. One did arise not long afterward, but there was absolutely none then. Even the dry tips of the lingering hedge-mustard, grey and blighted, and the fringe on the roof of the standing democrat-wagon were unstirred.

And yet amid that tense, godless calm the high bare boughs of all the trees in the yard were moving. They were twitching morbidly and spasmodically, clawing in convulsive and epileptic madness at the moonlit clouds; scratching impotently in the noxious air as if jerked by some alien and bodiless line of linkage with subterrene horrors writhing and struggling below the black roots.

Not a man breathed for several seconds. Then a cloud of darker depth passed over the moon, and the silhouette of clutching branches faded out momentarily. At this there was a general cry; muffled with awe, but husky and almost identical from every throat. For the terror had not faded with the silhouette, and in a fearsome instant of deeper darkness the watchers saw wriggling at that treetop height a thousand tiny points of faint and unhallowed radiance, tipping each bough like the fire of St. Elmo or the flames that came down on the apostles' heads at Pentecost. It was a monstrous constellation of unnatural light, like a glutted swarm of corpse-fed fireflies dancing hellish sarabands over an accursed marsh; and its colour was that same nameless intrusion which Ammi had come to recognise and dread. All the while the shaft of phosphorescence from the well was getting brighter and brighter, bringing to the minds of the huddled men a sense of doom and abnormality which far outraced any image their conscious minds could form. It was no longer *shining* out, it was *pouring* out; and as the shapeless stream of unplaceable colour left the well it seemed to flow directly into the sky.

The veterinary shivered, and walked to the front door to drop the heavy extra bar across it. Ammi shook no less, and had to tug and point for lack of a controllable voice when he wished to draw notice to the growing luminosity of the trees. The neighing and stamping of the horses had become utterly frightful, but not a soul of that group in the old house would have ventured forth for any earthly reward. With the moments the shining of the trees increased, while their restless branches seemed to strain more and more toward verticality. The wood of the well-sweep was shining now, and presently a policeman dumbly pointed to some wooden sheds and bee-hives near the stone wall on the west. They were commencing to shine, too, though the tethered vehicles of the visitors seemed so far unaffected. Then there was a wild commotion and clopping in the road, and as Ammi quenched the lamp for better seeing they realised that the span of frantic greys had broke their sapling and run off with the democrat-wagon.

The shock served to loosen several tongues, and embarrassed whispers were exchanged. "It spreads on everything organic that's been around here," muttered the medical examiner. No one replied, but the man who had been in the well gave a hint that his long pole must have stirred up something intangible. "It was awful," he added. "There was no bottom at all. Just ooze and bubbles and the feeling of something lurking under there." Ammi's horse still pawed and screamed deafeningly in the road outside, and nearly drowned its owner's faint quaver as he mumbled his formless reflections. "It come from that stone . . . it growed down thar . . . it got everything livin' . . . it fed itself on 'em, mind and body . . . Thad an' Mernie, Zenas an' Nabby . . . Nahum was the last . . . they all drunk the water . . . it got strong on 'em . . . it come from beyond, whar things ain't like they be here . . . now it's goin' home. . . ."

At this point, as the column of unknown colour flared suddenly stronger and began to weave itself into fantastic suggestions of shape which each spectator later described differently, there came from poor tethered Hero such a sound as no man before or since ever heard from a horse. Every person in that low-pitched sitting room stopped his ears, and

Ammi turned away from the window in horror and nausea. Words could not convey it—when Ammi looked out again the hapless beast lay huddled inert on the moonlit ground between the splintered shafts of the buggy. That was the last of Hero till they buried him next day. But the present was no time to mourn, for almost at this instant a detective silently called attention to something terrible in the very room with them. In the absence of the lamplight it was clear that a faint phosphorescence had begun to pervade the entire apartment. It glowed on the broad-planked floor and the fragment of rag carpet, and shimmered over the sashes of the small-paned windows. It ran up and down the exposed corner-posts, coruscated about the shelf and mantel, and infected the very doors and furniture. Each minute saw it strengthen, and at last it was very plain that healthy living things must leave that house.

Ammi shewed them the back door and the path up through the fields to the ten-acre pasture. They walked and stumbled as in a dream, and did not dare look back till they were far away on the high ground. They were glad of the path, for they could not have gone the front way, by that well. It was bad enough passing the glowing barn and sheds, and those shining orchard trees with their gnarled, fiendish contours; but thank heaven the branches did their worst twisting high up. The moon went under some very black clouds as they crossed the rustic bridge over Chapman's Brook, and it was blind groping from there to the open meadows.

When they looked back toward the valley and the distant Gardner place at the bottom they saw a fearsome sight. All the farm was shining with the hideous unknown blend of colour; trees, buildings, and even such grass and herbage as had not been wholly changed to lethal grey brittleness. The boughs were all straining skyward, tipped with tongues of foul flame, and lambent tricklings of the same monstrous fire were creeping about the ridgepoles of the house, barn, and sheds. It was a scene from a vision of Fuseli, and over all the rest reigned that riot of luminous amorphousness, that alien and undimensioned rainbow of cryptic poison from the well—seething, feeling, lapping, reaching, scintillating, straining, and malignly bubbling in its cosmic and unrecognisable chromaticism.

Then without warning the hideous thing shot vertically up toward the sky like a rocket or meteor, leaving behind no trail and disappearing through a round and curiously regular hole in the clouds before any man could gasp or cry out. No watcher can ever forget that sight, and Ammi stared blankly at the stars of Cygnus, Deneb twinkling above the others, where the unknown colour had melted into the Milky Way. But his gaze was the next moment called swiftly to earth by the crackling in the valley. It was just that. Only a wooden ripping and crackling, and not an explosion, as so many others of the party vowed. Yet the outcome was the same, for in one feverish, kaleidoscopic instant there burst up from that doomed and accursed farm a gleamingly eruptive cataclysm of unnatural sparks and substance; blurring the glance of the few who saw it, and sending forth to the zenith a bombarding cloudburst of such coloured and fantastic fragments as our universe must needs disown. Through quickly re-closing vapours they followed the great morbidity that had vanished, and in another second they had vanished too. Behind and below was only a darkness to which the men dared not return, and all about was a mounting wind which seemed to sweep down in black, frore gusts from interstellar space. It shrieked and howled, and lashed the fields and distorted woods in a mad cosmic frenzy, till soon the trembling party realised it would be no use waiting for the moon to shew what was left down there at Nahum's.

Too awed even to hint theories, the seven shaking men trudged back toward Arkham by the north road. Ammi was worse than his fellows, and begged them to see him inside his own kitchen, instead of keeping straight on to town. He did not wish to cross the nighted, wind-whipped woods alone to his home on the main road. For he had had an added shock that the others were spared, and was crushed forever with a brooding fear he dared not even mention for many years to come. As the rest of the watchers on that tempestuous hill had stolidly set their faces toward the road, Ammi had looked back an instant at the shadowed valley of desolation so lately sheltering his ill-starred friend. And from that stricken, far-away spot he had seen something feebly rise, only to sink down again upon the place from which the great shapeless horror had shot into the sky. It was just a colour—but not any colour of our earth or heavens. And because Ammi recognised that colour, and knew that this last faint remnant must still lurk down there in the well, he has never been quite right since.

Ammi would never go near the place again. It is over half a century now since the horror happened, but he has never been there, and will be glad when the new reservoir blots it out. I shall be glad, too, for I do not like the way the sunlight changed colour around the mouth of that abandoned well I passed. I hope the water will always be very deep—but even so, I shall never drink it. I do not think I shall visit the Arkham country hereafter. Three of the men who had been with Ammi returned the next morning to see the ruins by daylight, but there were not any real ruins. Only the bricks of the chimney, the stones of the cellar, some mineral and metallic litter here and there, and the rim of that nefandous well. Save for Ammi's dead horse, which they towed away and buried, and the buggy which they shortly returned to him, everything that had ever been living had gone. Five eldritch acres of dusty grey desert remained, nor has anything ever grown there since. To this day it sprawls open to the sky like a great spot eaten by acid in the woods and fields, and the few who have ever dared glimpse it in spite of the rural tales have named it "the blasted heath".

The rural tales are queer. They might be even queerer if city men and college chemists could be interested enough to analyse the water from that disused well, or the grey dust that no wind seems ever to disperse. Botanists, too, ought to study the stunted flora on the borders of that spot, for they might shed light on the country notion that the blight is spreading—little by little, perhaps an inch a year. People say the colour of the neighbouring herbage is not quite right in the spring, and that wild things leave queer prints in the light winter snow. Snow never seems quite so heavy on the blasted heath as it is elsewhere. Horses—the few that are left in this motor age—grow skittish in the silent valley; and hunters cannot depend on their dogs too near the splotch of greyish dust.

They say the mental influences are very bad, too. Numbers went queer in the years after Nahum's taking, and always they lacked the power to get away. Then the stronger-minded folk all left the region, and only the foreigners tried to live in the crumbling old homesteads. They could not stay, though; and one sometimes wonders what insight beyond ours their wild, weird stores of whispered magic have given them. Their dreams at night, they protest, are very horrible in that grotesque country; and surely the very look of the dark realm is enough to stir a morbid fancy. No traveller has ever escaped a sense of strangeness in those deep ravines, and artists shiver as they paint thick woods whose mystery is as much of the spirit as of the eye. I myself am curious about the sensation I derived from my one lone walk before Ammi told me his tale. When twilight came I had vaguely wished some clouds would gather, for an odd timidity about the deep skyey voids above had crept into my soul.

Do not ask me for my opinion. I do not know—that is all. There was no one but Ammi to question; for Arkham people will not talk about the strange days, and all three professors who saw the aërolite and its coloured globule are dead. There were other globules—depend upon that. One must have fed itself and escaped, and probably there was another which was too late. No doubt it is still down the well—I know there was something wrong with the sunlight I saw above that miasmal brink. The rustics say the blight creeps an inch a year, so perhaps there is a kind of growth or nourishment even now. But whatever daemon hatchling is there, it must be tethered to something or else it would quickly spread. Is it fastened to the roots of those trees that claw the air? One of the current Arkham tales is about fat oaks that shine and move as they ought not to do at night.

What it is, only God knows. In terms of matter I suppose the thing Ammi described would be called a gas, but this gas obeyed laws that are not of our cosmos. This was no fruit of such worlds and suns as shine on the telescopes and photographic plates of our observatories. This was no breath from the skies whose motions and dimensions our astronomers measure or deem too vast to measure. It was just a colour out of space—a frightful messenger from unformed realms of infinity beyond all Nature as we know it; from realms whose mere existence stuns the brain and numbs us with the black extra-cosmic gulfs it throws open before our frenzied eyes.

I doubt very much if Ammi consciously lied to me, and I do not think his tale was all a freak of madness as the townfolk had forewarned. Something terrible came to the hills and valleys on that meteor, and something terrible—though I know not in what proportion—still remains. I shall be glad to see the water come. Meanwhile I hope nothing will happen to Ammi. He saw so much of the thing—and its influence was so insidious. Why has he never been able to move away? How clearly he recalled those dying words of Nahum's—"can't git away . . . draws ye . . . ye know summ'at's comin', but 'tain't no use. . . ." Ammi is such a good old man—when the reservoir gang gets to work I must write the chief engineer to keep a sharp watch on him. I would hate to think of him as the grey, twisted, brittle monstrosity which persists more and more in troubling my sleep.

Howard Phillips Lovecraft,

August 20, 1890 – March 15, 1937)

Known as **H.P. Lovecraft**—was an American author who achieved posthumous fame through his influential works of horror fiction. Virtually unknown and only published in pulp magazines before he died in poverty, he is now regarded as one of the most significant 20th-century authors in his genre.

Lovecraft was born in Providence, Rhode Island. where he spent most of his life. His father was confined to a mental institution when Lovecraft was three years old. His grandfather, a wealthy businessman, enjoyed storytelling and was an early influence. Intellectually precocious but sensitive, Lovecraft began composing rudimentary horror tales by the age of eight, but suffered from overwhelming feelings of anxiety. He encountered problems with classmates in school, and was kept at home by his highly-strung and overbearing mother for illnesses that may have been psychosomatic. In high school, Lovecraft was able to better connect with his peers and form friendships. He also involved neighborhood children in elaborate make-believe projects, only regretfully ceasing the activity at seventeen years old. Despite leaving school in 1908 without graduating—he found mathematics particularly difficult—Lovecraft had developed a formidable knowledge of his favored subjects, such as history, linguistics, chemistry, and astronomy.

Although he seems to have had some social life, attending meetings of a club for local young men, Lovecraft, in early adulthood, was established in a reclusive 'nightbird' lifestyle without occupation or pursuit of romantic adventures. In 1913, his conduct of a long running controversy in the letters page of a story magazine led to his being invited to participate in an amateur journalism association. Encouraged, he started circulating his stories; he was 31 at the time of his first publication in a professional magazine. Lovecraft contracted a marriage to an older woman he had met at an association conference. By age 34, he was a regular contributor to newly founded Weird Tales; magazine; he turned down an offer of the editorship.

Lovecraft returned to Providence from New York in 1926, and over the next nine months, he produced some of his most celebrated tales including "The Call of Cthulhu", canonical to the Cthulhn Mythos. Never able to support himself from earnings as author and editor, Lovecraft saw commercial success increasingly elude him in this latter period, partly because he lacked the confidence and drive to promote himself. He subsisted in progressively straitened circumstances in his last years; an inheritance was completely spent by the time he died at the age of 46.

▼

August 20, 1890 - March 15, 1937

Age 9

Legal Stuff

Any likeness to those living or dead is merely coincidental, except where intended.

Nods

Earthpulse.com
Dr. Nick Begich and Jeane Manning
http://www.earthpulse.com/src/subcategory.asp?catid=1&subcatid=2

Earthsky.org
weatherwarefareblogspot.com
History Channel Documentary
Project Blue Beam

David Openheimer
http://www.bibliotecapleyades.net/sociopolitica/esp_sociopol_bluebeam04.htm

Cutting Edge
http://www.cuttingedge.org/news/n1806.cfm

EcoWatch
http://ecowatch.com/2014/11/27/government-spying-undermines-climate-action/

Documentary Heaven
http://documentaryheaven.com/the-invisible-machine-electromagnetic-warfare/

Lisa Clark

The Weekly World News: Chad Kultgen

Newsweek: John Hammer

Einstein's Theory of Relativity

John Wheeler: Annals of Physics

Andrew D. Basiago

John Titor

Larry Haber

Immortal Celebrities (*You know who you are!*)
Techblog.com
http://www.techeblog.com/index.php/tech-gadget/5-people-who-might-be-real-time-travelers

H.P. Lovecraft

. . . and all the other *wonderful unknowns* . . .

About The Author

She is fascinated by anything that has to do with the supernatural, the paranormal. This led her to become the author of ***Nightfall's Day*** and ***Blooded [Anunnaki Rising***], but she doesn't want you to confuse Blooded for just another young adult supernatural thriller with love triangles, vampires, werewolves, zombies, or dystopian societies. Blooded [Anunnaki Rising] blends the supernatural with what is perceived as mythological, historical fiction, [in which there may be more truth than fiction], while leaving readers considering the possibility that her spin on vampires might just be the real truth behind the legend.

A prolific writer of fiction, Lyn has more than a dozen books available for Kindle lovers, including two children's books in the "Little Book of Memories" series, which are also available in hard copy, and creative novella's that capture Lyn's diverse writing style, that include horror stories, stories filled with mystery and intrigue, ghost stories, love stories, and more.

An intuitive, hopeless romantic, Lyn loves science fiction, horror, and spirituality (but don't confuse that for being supportive of mainstream religion). Her family has a politically rich history and is tied to the American Revolution. When she's not playing World of Warcraft with her son, researching natural healing methods, or feeding the ducks in her lake, she's spinning tales of mysterious what-ifs for you.

A virtual recluse in her home, Villa Le Paradis Sur Terre, Lyn spends her days researching, reading, writing, and enjoying the simple things in life with her husband, such as a good cup of coffee and quiet conversation. The back of her Villa is glass from floor to ceiling and overlooks a private lake. I would say, that if ever there was truth in a statement, Lyn is living proof – that people who live in glass houses shouldn't throw stones . . . they should be writers.

You'll find Lyn Murray's Book Trailer on YouTube at the link below:

http://www.youtube.com/channel/UCxds7uuT4IMBRPLiYTNCGBQ?feature=watch

Lyn's Other Books

Little Book of Memories, Vol. 1
Little Book of Memories, Vol. 2
One Dark Halloween Night
The Howling Man
The Tuck
Nightfall's Day
Glasses Glasses
Paula [A Nightmare]
The 3rd Sunday of Every Month
[Mystery of White Rose Cemetery]
Who Goes There?
[The Legend of Tally Bottom Ridge]
A Case of Jitters
[Murder at Hammond Hill Rectory]
B.E.K. [Black Eyed Kid's Phenomenon]
Town At The End of Nowhere

♦♦♦ and of course ♦♦♦

Blooded [Anunnaki Rising]
Blooded – Nomads [Anunnaki Tribulation]
Blooded – Cinder [Anunnaki Armegeddon]
HUGENS
BANE
INDIGENOUS [Bigfoot People]
Weird Tales

. . . with more on the way . . .

◇◇◇

The Wilde Side
Peter Wilde Detective
Lyn's new detective series.

YOU'LL FIND LYN's BOOKS ON AMAZON.COM:

◇◇◇

. . . Golden Panda Publishing on Google . . .

◇◇◇

Just GOOGLE Lyn's website:

Lyn Murray Writes 2
http://lynmurray.wix.com/lynmurray

◇◇◇

Lyn's Book Trailers on YouTube
http://www.youtube.com/channel/UCxds7uuT4IMBRPLiYTNCGBQ

OR

Just go to YouTube and type in:

Lyn Murray

(Find her book cover picture to locate the right channel.)

Thank You

I am honored that you took the time to read my book, and really hope you liked this anthology! I loved putting it together for you. If you could, take a moment to let me know what you liked about it, I'd really like to know. Your feedback helps me hone my skills. I'm always looking for new ideas, and developing characters and story plots. Tell me what kind of stories you like – I write for you!

See ya soon.

Bye!

I love you my God (my sweet Jesus) – I see you!

Our Intrepid Duo

Yesterday

Joe at 16

Who grew up to be a
Chrysler Corporation Executive,
Lee Iacocca Chairman's Award Recipient,
Researcher, Advocate, Activist, Author, Artist,
Poet, and Poker Champ.

Lyn at 16

Who grew up to be a
Film Processing Industry Executive,
Entrepreneur, Advocate, Activist, Author, Artist,
Poet Laureate, and World of Warcraft nerd.

888

A Prayer for Humanity

*We place our faith in you Father (the ALL) – Our Lord Jesus the Christ – the fate of mankind (the fate of all Your Creations – wherever they are) are in your hands. Take care of us Father – of me and mine, of all the **good people** of Earth and throughout the universe and beyond – especially the least of the least (be they plant, animal, human or otherwise) keep us from harm and from those who would do us harm! No matter what anyone says – with all that we are (with our hearts and souls) we trust you . . . **and we know that you ARE there, and that you will make right the wrong!** It is my prayer that all who seek You – find You . . . the REAL YOU – and that they see and renounce the "psychotic Fallen Angel gods" of Earth's religions!*

I Love You!

Thank You for loving me . . . and my family of man!